Applied Crowd Science

"Dr. Still was able to analyse the planning and layout of the concert and determine that the concert promoter had deviated from the life safety plan without notifying the proper authorities. Shortly after receiving Dr. Still's report, the defendants elected to settle the case."

Raymond R. Dieppa, FLORIDA LEGAL, LLC.

"Professor Still is the world's leading expert in crowd science. His first book on the subject is the foundation for an entire industry. This book develops his work even further and by using the contents and frameworks contained within this latest fantastic piece of work, organisers around the world will make their mass gatherings safer. For all crowd science and crowd safety practitioners, make sure this book is top of your reading list as we move into a post-Covid world."

D. Smyth, Founder and CEO of Sword Group

"I needed someone who could teach the higher echelon of the Indian Police Service about the science behind crowd management. . . . Thankfully, Professor Keith Still agreed to impart his phenomenal knowledge on dealing with crowds to the course and his presentation has since become a key feature. It is safe to say that Keith is one of, if not the world leader, in defining the considerations relating to potential crowded places, planned or otherwise. One can only speculate as to the thousands of lives that have been saved due to Keith's work and not one of those individuals realises."

Graham M. Sunderland, QPM, MA

"The danger for police commanders in planning and dealing with major events is that we become focussed on one area, normally public order, when in fact we should be considering a range of overlapping and inter-related disciplines of which crowd science should be writ large. An awareness of the area should be a prerequisite for those policing events and commanders should be able to recognise risk at events and utilise the available specialists in appropriate circumstances. I can think of one high-profile football match . . . where I was glad issues of crowd science were in my considerations as there was clear potential for a serious incident if the wrong calls had been made. Consequently, in planning for this year's Manchester United victory parade, one of my first calls was to Keith Still to ensure I had his support in reviewing the issues around crowd science – a previous parade had attracted 250,000 fans and there had been a real risk of people getting seriously hurt. In this case, Keith's involvement gave me the reassurance that the issues would be recognised and that we could develop the right plans and mitigation measures."

Mark Roberts, Chief Superintendent, Trafford Division,
Greater Manchester Police (2011)

Applied Crowd Science

G. Keith Still BSC PHD FIMA FICPEM FIPM FHEA MEA

Artwork and graphics by
Erin Still

CRC Press
Taylor & Francis Group
Boca Raton London New York

CRC Press is an imprint of the
Taylor & Francis Group, an **informa** business

First edition published 2022
by CRC Press
6000 Broken Sound Parkway NW, Suite 300, Boca Raton, FL 33487-2742

and by CRC Press
2 Park Square, Milton Park, Abingdon, Oxon, OX14 4RN

CRC Press is an imprint of Taylor & Francis Group, LLC

Library of Congress Cataloging-in-Publication Data
Names: Still, G. Keith, author.
Title: Applied crowd science / G. Keith Still.
Description: First edition. | Boca Raton, FL : CRC Press, 2022. |
Includes bibliographical references and index.
Identifiers: LCCN 2021035046 (print) | LCCN 2021035047 (ebook) |
ISBN 9781138626560 (hardback) | ISBN 9781351053068 (ebook)
Subjects: LCSH: Crowds. | Crowd control. | Risk assessment.
Classification: LCC HM871 .S74 2022 (print) | LCC HM871 (ebook) |
DDC 302.33—dc23
LC record available at https://lccn.loc.gov/2021035046
LC ebook record available at https://lccn.loc.gov/2021035047

ISBN: 978-1-138-62656-0 (hbk)
ISBN: 978-1-032-17083-1 (pbk)
ISBN: 978-1-351-05306-8 (ebk)

DOI: 10.1201/9781351053068

Typeset in Sabon
by codeMantra

Access the Support Material: www.routledge.com/9781138626560

To my wife, Valerie, none of this would be possible without your love and support, and to Harry and Erin, who are my motivation to try and make the world a safer place.

Contents

Opening remarks

Like many of you reading this opening paragraph, I have been a devotee of the work of Keith for many years. I've also had the pleasure and continue to have the pleasure of knowing Keith personally and of having him on my speed dial whenever I'm needing sage advice in any of the multitude of areas of expertise which he holds within his vast range of competence.

For many years through his teachings and publications Keith has offered support to many of us helping us all to broaden our horizons, challenge with confidence and think in ways that perhaps we would not have done before meeting him.

This latest book *Applied Crowd Science* which follows neatly on from an earlier publication *Introduction to Crowd Science* reminds technicians like me that it's not sufficient to simply understand the geometry of a building or an event, but that we must also understand the criticality of, and linkage to the psychology of human behaviour which will manifest from those intending to occupy the space.

The fascination for me, and appreciation is in the way that work in this field has moved on from the prediction of crowd behaviour to now looking at ways in which that behaviour can be shaped in positive ways, with positive messages all of which will add to the maintaining of high levels of safety and customer experience.

This book will act as a useful tool and aide memoire for those undertaking or about to undertake the Level 3 course and benefits enormously by its simplicity of layout, the linkage to the course itself, and for me, the icing on the cake being the harvesting of the case studies, carried out by graduates of the course. Drawing on of real-life examples brings to life the book and importantly gives a platform to those graduates on which they too can showcase their emerging talents to an audience of their peers.

I wish my friend Keith and all the contributors every success with this latest publication and look forward to reading and being educated by many more in years to come.

Ken Scott MBE FRICS
Acting Chief Executive.
The Sports Grounds Safety Authority
June 2021

About the Author

> Safety is not an intellectual exercise to keep us in work. It is a matter of life and death. It is the sum of our contributions to safety management that determines whether the people we work with live or die.
>
> *Sir Brian Appleton after the Piper Alpha Disaster*

Prof. Dr G. Keith Still is a Fellow of the Institute of Mathematics and its Applications (FIMA), a fellow of the Institute of Civil Protection and Emergency Management (FICPEM) and a Specialist Fellow of the International Institute of Risk and Safety Management (SFIIRSM) a Fellow of the Institute of Place Management and a Fellow of the Higher Education Academy. He was a member of the Expert Witness Institute until 2021 when he became a Member of the Academy of Experts.

Keith is a visiting Professor of Crowd Science at the University of Suffolk, UK. He teaches the principles and applications of crowd science (crowd safety, crowd dynamics, crowd risk analysis) around the world. Keith lectured at the UK Cabinet Office Emergency Planning College (EPC) from 1999 to 2014. He developed and taught short courses for crowd safety in places of public assembly and contributed to several other safety-related courses at the EPC.

His mathematical and human behaviour modelling led to the coding and development of several prototype crowd simulations, such as Legion (his PhD thesis was the creation of the Legion system), Paramics UAF, Myriad, Myriad 3D, Shepard, 3DTRDS, Tawaf, Dwell plus several other crowd models.

His simulation tools have led to the development of a systematic blueprint for crowd modelling aspects of crowd safety and crowd risk analysis in the built and complex environment. Developing and using these tools, Keith has consulted and worked as a safety coordinator, safety advisor, safety manager, strategic, tactical and design advisor on many of the world's most dangerous and challenging events. Projects include the Mega-events such as the Olympics, the annual Pilgrimage to Makkah (Hajj), Canada Day, London, and Dubai New Year events etc. Keith developed, used, then sold the rights to several of his prototype computer simulations to fund the development of his teaching and training programmes.

Although he has developed and used some of the most sophisticated crowd modelling and simulation tools, his work remains true to the principles of his original research: modelling crowds and crowd dynamics. His work is directed towards improving crowd safety and crowd risk analysis in places of public assembly.

To this aim, his teaching and training programmes do not use computer simulations. He focuses on applying of risk analysis methods for front line staff, management, planners, and safety advisors for large scale events. He has taught well over 10,000 delegates over the last few decades, both through face-to-face classroom courses and through his eLearning platform (GKSed.com and CrowdRisks.com).

In 2020, due to the Covid-19 Pandemic, all his courses moved online and over 1,000 learners passed through his crowd science courses in 2020–2021, the lockdown year (s).

This book follows the online content, with example essays and case studies for applied crowd science. Keith covers the VLE (Virtual Learning Environment), how to access the course, the course content, and example essays and case studies.

It was written to provide a deeper insight to the development, the applications and the theories that underpin his research and educational materials.

Foreword from Ken Scott MBE FRICS

In the UK we are well-practised in being part of a crowd and for our skills and patience in forming queues which are often the product of those crowds. A significant portion of this is from experiences of being part of a crowd at a sporting event. Indeed, there is an old adage that if queueing was an Olympic sport, Team GB would win the gold medal every time.

Sport is an essential part of our national culture. Every week millions of people in the UK attend football, rugby, horse racing and other sporting events. Thanks to the dedication and commitment of all those who work in stadium safety, this country is one of the safest places in the world to watch live sport. But how has this come about?

For anyone involved in the management, organisation or hosting of an event, the safety of all people present must take precedence over every other concern. Safety at sports grounds is achieved by establishing a balance between good design, good information systems and good management. It is useful in this respect to consider the three central tenets of safety, security and service.

Safety identifies the importance of overlay around a sports ground or venue, which can impact the safe arrival, circulation and egress of a crowd and considerations must be given when planning any security measures to ensure that safety is not compromised.

At times there is also a need to consider additional security overlays to manage risks at an event, whether it is enhanced screening/searching, vehicle barriers, and so on. At the same time, consideration must also be given to understanding the unintended consequences of these, for example, increased queues and delayed entry into a sports ground.

Furthermore, there is a need to balance safety and security against customer experience, the service element, and area which, sadly, is often overlooked. These measures are aimed at the well-being of all people at an event. Experience shows that the behaviour of people at an event can be significantly influenced by conditions at the venue, and by the attitudes and actions of the event staff.

Crowd safety cannot be achieved simply by ensuring that individual components of a ground or venue – such as stairways, gangways, seated areas, or terraces – are satisfactory in themselves. The inter-relation of these and other components – such as concourses and CCTV systems – is critical. None can be treated in isolation without consideration of the effect its design and management have upon other components. They should all be compatible and combine to form a balanced unit. Furthermore, good safety management will not necessarily compensate for poor design, or vice versa.

The Sports Grounds Safety Authority (SGSA) has led the way on this understanding by identifying that consideration must be given to both the physical and safety management factors when it comes to calculating a safe capacity – something referred to in the Guide to Safety at Sports Grounds (also known as the Green Guide) as (P) and (S) Factors, respectively.

But this concept, like so many crowd management strategies, infrastructure and procedures that are taken for granted today, has taken time to establish and embed over many years. Indeed, much of the work that makes the UK one of the safest places in the world to watch and enjoy live sport and entertainment has been borne out of tragedy. In the 1970s and 1980s several tragedies led to the changing of infrastructure and procedures and the introduction of legislation, which have led to the safety record the UK has today.

The catalyst for the changes was the disaster at Ibrox Park in Glasgow in 1971. On that tragic January day Rangers were playing Celtic in the Old Firm derby 66 people lost their lives and over 200 injured. The findings of the Inquiry into the disaster, led by Lord Wheatley, were the basis of the first edition of the Guide to Safety at Sports Grounds – the Green Guide – in 1973.

Just over a decade later in 1985, the country suffered again when 56 people died and hundreds more were injured at Bradford when a timber stand caught fire. It was one of the worst stadium fire disasters in history.

The 1980s were also synonymous with football hooliganism in the UK and just 3 weeks after the Bradford tragedy in May 1985 there was another disaster at the European Cup Final between Juventus and Liverpool in Heysel Stadium, Brussels which resulted in the deaths of 39 people with more than 600 injured.

But still the disasters kept coming and on 15 April 1989, in one of the worst ever international football disasters, 96 people lost their lives and many hundreds more were injured at Hillsborough Sheffield. This tragedy was the culmination of a series of factors and is forever etched on the nation's collective memory. It led to an Inquiry led by Lord Justice Taylor. The findings of this enquiry led to many of the changes that are in effect today, including dedicated safety officers at grounds and all-seated stadia for clubs in the top two tiers of football in England and Wales. As stated by Lord Taylor's report: "complacency is the enemy of safety". This is something that anyone involved in crowd management should be mindful of.

The Hillsborough tragedy was a turning point for sports ground safety and crowd management in the UK. Over the last 30 years there has been a step change to develop safe, modern stadia with professional safety teams in place with in-depth crowd management knowledge and experience. The effect of this has been no deaths at sports grounds because of safety management failures since Hillsborough.

Of course, that does not mean that safety issues are now a thing of the past. The challenges we face today are not the same as those as the 1970s and 1980s. There is a risk attached to every event. The only completely safe stadium is an empty one...but during the COVID-19 pandemic in 2020 we saw enough of that, and nobody wants a repeat of that situation. The last decade has made clear the changing environment and risks for crowd safety.

The terrorist incidents which have occurred, including Stade de France in 2015 and the attacks at Westminster Bridge, Manchester Arena and London Bridge in 2017 are a stark reminder about the vulnerability of crowded places and the importance of the work of those charged with keeping us safe. Sadly, in today's world, we now need to consider the increasing threat posed by such attacks when we are developing plans. This was at the forefront of our minds when developing the sixth edition of the Green Guide, published in 2018. The document broke new ground in recognising the crucial part that areas outside of the ground play in the safe arrival and departure of spectators – referred to as Zone Ex. The concept has been further expanded upon in the SGSA's Supplementary Guidance 03: Event Safety Management.

Stadium and event operators can no longer consider the space outside the stadium as a responsibility of others. These areas form crucial elements in the safe arrival and departure of spectators and need full and open discussion with all partners to develop plans for their safe management.

When considering safety, the sum of all parts is critical to creating as safe an environment as possible for everyone at the ground. There are many challenges to ensuring safety in the area, which is Zone Ex, not least of which is that it is unlikely that the venue owner has any legal responsibility for people within it. A multi-disciplinary approach is needed, working across disciplines such as safety, police, stadium management, ambulance, and other key stakeholders to ensure that someone has the lead for safety in this area.

The boundaries of a venue, as defined on a plan or a safety certificate, are often the boundary at which legal responsibility for spectators starts and ends for the venue operators. However, there is a moral obligation to ensure that consideration is given to the area beyond this line (which exists only on paper and is not painted on the floor). While this area may not be the direct responsibility of the stadium owner, it is important that all parties are involved in the effective management of this zone to ensure that spectators are safe during ingress and egress.

Advances in crowd modelling can help plan for these situations. In stadium design, it is important to understand people movement; using modelling and analysis, we no longer need to wait for the venue to open to realise what does and does not work. We can get a good idea of outcomes from using technology and simulating how people move around sports grounds under certain conditions. The technology that can help us assess that is improving all the time and is a powerful tool in stadium design and crowd management.

Similarly, all safety managers need to be aware of the options they have available in their toolbox to help tackle any challenge. This is not about writing countless plans for every emergency possible. At times of emergency, you do not have the time to consider if the most appropriate plan is number 536 or 537. Instead, the most useful piece of preparatory work a safety manager can do is to develop a complete understanding of all the options and tools available that can be used in a difficult situation.

What can never be underestimated, of course, is the importance of effective engagement with crowds when it comes to safety. Effective crowd management needs to involve an awareness of how engaging with crowds can benefit the safety of the event overall. For example, a sports ground or event venue may want to consider the whole spectator journey – from when a spectator leaves their house, to ingress arrangements to a safe egress. This can include sharing safety messages before the game through apps and social media, the experience of fans travelling via public transport and entering the leaving the ground through Zone Ex, which is likely to encompass the main pedestrian and vehicle routes.

The importance of effectively engaging and working with fans is embedded within the Council of Europe Convention. This is the only internationally binding instrument to establish an integrated approach based on safety, security and service. It also promotes co-operation between all public and private stakeholders: governments, municipal authorities, police, football authorities and supporters, to provide a safe, secure and welcoming environment at football matches and other events, whether inside or outside the stadium.

Engaging with fans early on can help to reduce some of the unintended consequences of any increased safety and/or security requirements. This could be providing additional entertainment inside the ground to encourage people to arrive early. The days of someone turning up to a football match 15 minutes before kick-off and expect to be in immediately are a thing of the past, so grounds and venues need to create "incentives" to encourage early arrivals.

Whilst technology has its part to play, we should never lose sight of the important role of venue staff can play in overall safety equation. The most important asset a sports ground has in terms of safety is the staff. Effective and vigilant staff, who provide excellent customer service to those attending an event are a critical element to a safe, enjoyable experience for fans.

The nature and scale of issues affecting crowd safety today is significant and fast-changing. It is vital that the sector continues to look ahead, to adapt and respond to the changing world to ensure millions of people can enjoy going to sports grounds and venues in safety.

There is a risk attached to every event. As already said, the only safe stadium is an empty one. But there are ways to manage and mitigate risk effectively to create as safe an environment for crowds as possible. However, we are careful not to rest on laurels and recognise that the moment we close our minds to what is happening out there is the time we become complacent. To heed the words of Lord Justice Taylor, those involved in crowd management must always guard against complacency in safety.

Finally, I could not write this Foreword without reference to the recent Covid pandemic but will not major on its effects which have wreaked upon our sector which are all well documented.

I will though take a moment to spell out the need to consider and factor in human factors and behavioural considerations into safety, security and service considerations. Sadly, these important considerations do not always receive sufficient thought and attention.

Whether society ever reaches the point of a "post-covid" world remains to be seen. But as we all emerge from various lockdowns around the world into a largely vaccinated world, the need to consider those human factors and likely behavioural patterns has never been more relevant and important to those planning to host "safe" events.

The simple instructions around social distancing and avoiding crowded spaces are now ingrained within the mind set of many and will be a habit which is hard to break. There is a saying that it takes 6 weeks to form a habit so due to the length of pandemic restrictions we are all well beyond that point.

Behaviours as we all as a society emerge into the new Covid aware world will be impacted upon by many different considerations such as, but not exclusively, personal experience during the peaks of the pandemic, age, gender, health vulnerabilities and personal risk appetite.

Confidence building within ourselves to return to the events that we all love will be key to getting back to "normal" whatever that may look like in a Covid world. That confidence has got to be built on clear, concise, and evidenced data, trust in the skills and competencies of those designing and managing events, high levels of customer care in delivering those events but importantly on a need for us all take responsibility for our own safety and taking an interest in the safety of others within our groups.

How long it will take to restore confidences will vary widely from individual to individual, from event to event, and after consideration of my points above. But it is something that we must all strive to achieve to overcome the ravages of recent months and point to a better future for us all.

Ken Scott MBE FRICS
Acting Chief Executive.
The Sports Grounds Safety Authority
June 2021

Preface

> Concern for man himself and his safety must always for the chief interest of all technical endeavours.
>
> Albert Einstein

In our previous book, *Introduction to Crowd Science*, we outlined the background to a modelling approach for crowd safety and risk analysis for major events. There was an emphasis on not relying on the outputs of computer simulations to prove a crowded space is safe. We also outlined where computer simulations should be used. For example, to understand the crowd dynamics in a progressive crowd collapse, where it is too dangerous to assess risks with live subjects.

In *Introduction to Crowd Science*, we outlined the principles of crowd density, crowd flow and the risks of overcrowding using the five risk modelling tools. These are DIM-ICE model, Crowd dynamics (crowd density and crowd flow rates), RAMP Analysis, Risk and Congestion Mapping and the Decision Support Analysis tools. Together these provide the user with a set of tools. Those tools provide an insight of the dynamics of crowd risks in places of public assembly.

The DIM-ICE risk analysis model is now integrated in the Events Guide[1] crowd management (section 13) and the Guide to Safety at Sports Grounds[2] (density and flow). We are working with the National Center for Spectator Sports Safety and Security training programmes at The University of Southern Mississippi (USA) with a USA version of the Level 3 course.

Since the publication of *Introduction to Crowd Science*, the suite of crowd risk analysis tools was adopted by the UK College of Policing, into the Public Event Commanders Course (the PEC's course) and now (June 2021) the New York Police Department (USA) and the Police in New South Wales (Australia) are working with us to develop their own versions of the PEC's course.

[1] https://www.thepurpleguide.co.uk
[2] https://sgsa.org.uk/greenguide/

After the publication of *Introduction to Crowd Science*, we had several requests to structure a book *around* the courses[3] we teach, to provide example essays and case studies, and to update the materials for both the USA and the UK guidance and legislation framework.

We continue to teach the *Introduction to Crowd Science* course, both online and in classrooms around the world. The course is taught at the Breda University of Applied Science. Elements of our work are taught around the world in various authorised (and many unauthorised) courses.

In this book, *Applied Crowd Science*, we expand on the description and application of the modelling tools, following the themes and structure of our training courses. We illustrate their applications for event safety and risk analysis in delegates and student case studies, written by graduates of our courses and the MSc course we developed and taught at Manchester Metropolitan University. This book draws on decades of teaching, both in a face to face (classroom), and the experience of converting these courses into an e-learning environment. Our courses run from Level 3 through the Level 5, the MSc in Crowd Safety and Risk Analysis. They are all self-paced and we are continuing to develop and enhance the materials.

This book is divided into three sections. The first relates to the core risk analysis modelling tools as taught on the Level 3 course. The second on the delegates crowd science essays, and the third section is case studies, from client sites, using the tools, illustrated with practical applications of the suite of crowd risk analysis tools. Due to the number of colour images and graphics in the case studies, we decided these are best suited for an online version of this book. The relevant links for the additional case studies: www.routledge.com/9781138626560.

By the end of this book, you will have a grounding in the theory, and application, of crowd science. As the saying goes, knowledge is useless, unless it is passed on. We hope you enjoy the journey.

[3] www.CrowdRisks.com

Acknowledgements

During the creation and production of a book, there are many people to thank. Firstly, thank you to all the authors who have contributed papers (we let them introduce themselves in their sections). In alphabetical order: Aldo Raineri (Australia), Andrew Seeds (UK), Bert Bruyninckx (Belgium), Breda University of Applied Science (Holland), Diarmuid Smyth (Ireland), Eric Kant (Holland), Graeme Clarke (Australia), Ise Murphy (Ireland), Ken Scott (UK), Kevin Kennedy (Canada), Kristin Finn (UK), Manon Gijbels (Holland), Marc Tubee (Holland), Mark Breen (Ireland), NCS^4 (USA), Roderick van Gelder (Australia), Ryoki Murayama (Japan), Shane Harris (Australia), Sheila Russell (UK), Steve Riches (UK), Syan Schaap (Holland), Tristan Taylor (Australia), Willem van den Berge (Holland). It was a pleasure collaborating with you all, and a huge thank you for your contributions, and patience.

Some of the above contributions are in full colour in the support folders. The print copy choice was not a reflection on the quality of the work. We were limited in the print costs and colour/greyscale conversions – some worked much better in colour.

To my daughter, Erin, who has supported all the work we have developed over the last 3 to 4 years. She has created all the artwork for my lectures and this book. I could not do as much, or for as long, without your tireless support. Erin, enormous thanks for all your hard work and assistance during the online workshops and courses. Thanks for always being there when I need something researched, illustrated, or checked.

Thanks to University of Suffolk, to NCS^4 (University of Southern Mississippi), Breda University of Applied Science BUAS), now in our third year of running the Crowd Safety Minor. We feature two of the BUAS student's work (winner and runner up, of the 2021 Student of the Year awards). Thanks to Charles Sturt University (John Dines and Graham Sutherland) for supporting my work in India on your MCTP III Police programmes, it is great to be part of that team. Thanks to Dr. Syan Schaap, and all at the Event Safety Institute (ESI), for all your support in delivering the course in Holland and Belgium, and for the support in the work at Breda University of Applied Science. Thanks to Phil and Jamie of the UK College of Policing, it was great working with you on the PEC's development. Thanks to Chris Kemp, for proof reading and edits of the student papers during the early drafts.

Thanks to Ken Scott, who I have enjoyed many long and detailed conversations on crowd safety, the Green Guide is a huge testimony to Ken's contribution to crowd safety and the Safety at Sports Grounds Guidance is guidance that, I know, is used around the world.

Huge thanks to Frazer Merritt of Taylor Francis for all his patience and assistance during the process of producing this book.

Finally, thank you to everyone who has supported, adopted, and used the risk analysis methods, we define in the courses, and in *Introduction to Crowd Science*. Thank you for all the positive feedback we have had over the last 30 years working towards improving crowd safety, around the world.

SPELLING CONVENTIONS

When editing papers from UK, USA, Canada, Europe, and Australia, we have standardised on using the English convention of "ise" instead of "ize" throughout this book. This is only for consistency in our grammar checking software and typesetting process.

ESSAYS AND CASE STUDIES

To include colour images and diagrams, there are further examples of the essays and case studies on the website. A complete grey scale and ***colour*** version of the book is available in eBook format.

Introduction

> Wisdom comes from experience. Experience is often a result of a lack of wisdom.
>
> Terry Pratchett

The MSc (Crowd safety and risk analysis) was taught 100% online and ran for several years with close to 100 students progressing through the Post Graduate Certificate (first-year pass), to the Post Graduate Diploma (second-year pass) and on to the MSc (year three pass, Dissertation by research). Most of the students who progressed through the course began their journey on one of our workshops, which ran from 1999, at the UK Cabinet Office Emergency Planning College, through to the Foundation Degree at Bucks New University and then to Manchester Metropolitan University. It was an absolute pleasure assisting in the progression of these Master Students, and several are still working with us on international projects.

As stated above, we shifted all the courses online during the pandemic and developed the VLE (Virtual Learning Environment) using Moodle. We created an online training course for the Institute of Place Management, *Covid-19: Place Management* for the business improvement districts (BID's) and the local high streets and city centres. While writing this introduction (June 2021), we anxiously await the news for the lockdown to end and a return to the time when the crowds can come back.

Before you start this book or any of our courses, there are some fundamentals components and terms that we should define.

WHAT IS CROWD SCIENCE?

Crowd science defines the how, where, when and why crowds form and move. There are elements of the geometry of crowded spaces (area, distance, capacity), the crowd dynamics (crowd density, crowd flow rates) and aspects of human psychology (individual and group behaviour). The latter includes how people react to each other, to their environment and to the cultural norms of their whereabouts.

Crowd science focuses on crowd safety and crowd risk analysis, in shaping crowd behaviour and expectations of public gatherings, and the safe monitoring and management of the crowd, for both normal and emergency situations. In this book, we cover the applications of the crowd sciences, to planning and managing places of public assembly.

MODELS AND SIMULATIONS

There have been significant improvements in crowd modelling using simulations, over the last decade. We have developed progressive crowd collapse[4] models for the Premier League and the Safety at Sports Grounds Authority[5] (SGSA). Simulations provide a safe testing environment, where you can evaluate the limits of safety, but you can never prove a system is *safe* for a crowd. There will always some level of risk associated with crowded spaces. The only *absolutely safe* environment is where there are no crowds, no performers, no staff and no event.

WHY IS THIS IMPORTANT?

During our work, in the development of crowd simulations, it was clear that no matter how good the models looked or how realistically the simulations modelled crowd movements, any changes in the environment could induce an unexpected crowd reaction. Specifically, any information changes in the environment could cause a person to change their course and hence affect the overall crowd dynamic.

For example, a person may change direction and/or behaviour due to real (or perceived) threats. Sudden changes in the weather, loud noises, strange/unpleasant smells, inconsistent signage, unexpected announcements, unexpected phone call, something their friend/partner says can all affect crowd behaviour. There are simply too many variables to list. That is the enigma facing all crowd simulations. How do you anticipate, model, and calibrate simulations to include all the potential variables and all the subsequent possible human reactions?

In computer simulations, the best you can hope for is to produce a result based on a specific set of initial inputs, and/or variables. In other words, you could never fully predict how a crowd may react to the many potential

4 Crash test dummies - for crowds.

5 https://sgsa.org.uk/wp-content/uploads/2021/06/The-management-of-persistent-standing-Final-report.pdf

unknown future influences. There are too many unknowns and too many variables that can change an outcome. Both at an individual level, and how that may change the overall crowd behaviour, reaction and crowd interactions during a mass gathering. Although this may be self-evident, it is demonstrable when you review a simulation output and ask the programmer, what happens if...

> ... *the weather changes.*
> ... *there is a sudden loud noise*
> ... *there is a suspect package, or a bomb, or an active shooter, or a fire, or* ...

How will your crowd reast within your simulation? Each change in the "what if ..." inputs may require a new simulation run and may produce a different outcome. This quickly becomes a vast array of potential "what ifs..." and further simulations run to evaluate all the possible conditions. The problem is the analytical intractability[6], the number of potential input variables, there are simply too many unknowns.

ASSESSING THE LIMITS

Although predicting individual behaviour, and hence, crowd behaviour may be a paradox of the unknown, you can evaluate certain boundary conditions, understand limits, and the overall crowd risks. Computer simulations can allow the experienced user to explore, in perfect safety, how systems may fail. You can experiment with the input parameters that might influence the crowd dynamics or the crowd behaviour to gain insight to the cause and effect. We may never know, for certain, the outcome of all the variables that might affect crowd behaviour, but we can explore the boundaries, to understand and assess the limits of the crowd dynamics. However, the questions may remain unanswered by simulations alone. We may never be able to calibrate all the behavioural aspects of a crowd.

INFLUENCING CROWD BEHAVIOUR

However, it *is* possible to influence crowd behaviour. You only need to watch a stage magician perform an illusion to understand how a crowd's expectations are managed, how perceptions can be shaped, influenced, and manipulated. What you see, what you think you see, what you thought you saw, these are all part of the carefully crafted art of the performance and

6 "Identifying Sources of Intractability in Cognitive Models: An Illustration using Analogical Structure Mapping" by Iris van Rooij.

illusion. Even the simplest of close-up magic illusions, making a coin vanish and then reappear, will influence your senses into following a chain of events, and manage your expectations.

In the paper, the misdirected quest[7], Lamont says the following:

> What history also shows is that experimental methods do not compensate for a superficial understanding of an enormous range of situational practices, and that even the most up-to-date technology does not necessarily provide better knowledge in matters of complex human interaction. After all, so far as psychology is about understanding and predicting human behaviour, the experienced magician can demonstrate his or her ability to do so regularly (with a replication rate most psychologists would envy), and to do so in the real world. Magicians have been able to do this for centuries, but how they have done this has changed according to the way in which people think and behave at different times and in different places. Psychologists can indeed learn from magic, but it is magicians who understand the real secrets.

FORM AND MOVE

As the elements of "why crowds form and move" are more grounded in the psychology of human behaviour, understanding how to direct attention, specifically how a stage magician[8] can influence a mass audience, proves us with a deep insight on the mechanisms of focussing the crowd's attention on the information[9] they present, and on how they influence their audiences behaviour. This became a natural extension of our research into crowd behaviour.

DIRECTING ATTENTION

The key question here is, rather than try and predict crowd behaviour, what would we need to shape, to influence and then control crowd behaviour? Would it be possible to combine the mathematics (crowd density, crowd flow, physical limitations of area, flow rates and crowd movement) with the same techniques used in audience management on the stage? Of course, we do not call this magic, but we do borrow from the same techniques used by the magician to influence their audience, for crowd safety applications.

7 https://thepsychologist.bps.org.uk/volume-23/edition-12/misdirected-quest

8 The author has studied, practiced, and performed close up and parlour illusions since he was a child. His Great Uncle (The Great Irvano) was a stage magician and sparked an early interest and lifelong passion for illusions, both small and large. He has an extensive library of rare magic books and original effects.

9 Signage, PA announcements, social media, wayfinding, etc.

IT IS NOT NEW SCIENCE

This is not new science or even a new application of the management of the crowds' expectations. Marketing companies[10] use these same techniques for product placement and sales campaigns. We have used those techniques to shape the crowds' expectations, influence their behaviour and structure a safer environment.

BACKGROUND REFERENCES

Figure 1 is part of the author's library, mathematics and magic (top shelf, bookcase 4 of 12).

In the "The Fitzkee Trilogy" first published in 1945, by Dariel Fitzkee[11], *Magic by Misdirection* applies the psychology of deception to the art of magic. Its central thesis is the principle of using a range of psychological devices and methods you can use to deceive your audience, such as

> Disguise, Attention Control, Suggestion, Simulation, Dissimulation, Interpretation, Manoeuvre, Pretence, Ruse, Anticipation, Diversion, Monotony, Premature Consummation, Confusion,

Figure 1 Part of our extensive reference library.

[10] https://www.amazon.co.uk/Why-We-Buy-Shopping-Internet/dp/B0073UQ0WW/ref=pd_lpo_card_2?pd_rd_i=B0073UQ0WW&psc=1

[11] https://www.vanishingincmagic.com/magic-theory/the-fitzkee-trilogy/

In the third book, *Showmanship for Magicians* Fitzkee outlines,

> ...the subject of presenting of magic to gain the greatest audience response, including a detailed analysis of spectator appeals and how to incorporate them into the presentation of magic.

Within this, many other books from 80 to 100 years ago on similar subjects are outlines of audience management, of shaping expectations. This was the era of the stage performer, creating illusions without the aid of cameras, electronics or any of the special effects possible today. An era where they had to master the art of communication with their audience, to create and sustain their illusions.

Banachek[12] in his books, *Psychological Subtleties* (1, 2 and 3) outlines a wide range of psychologic forces using subtleties in his use of words, his structuring of language, his positioning of objects, to create the illusions of free choice. In Andy Nyman's *Bulletproof*[13], on the "bonus tips on Killer Elite", he outlines the use of subliminal force, again to give the illusion of free choice.

Derren Brown's[14] books *Pure Effect* and *Absolute Magic,* offer a detailed study of how to make magic convincing, by weaving a persuasive narrative into the effects he performs, on how to shape the audience's perception of reality. He weaves information into the environment of several of his effects, to give the illusion of free choice, although he is in complete control of the outcome.

In R. Paul Wilson's book, *The Art of the Con* he details a wide range of scams and confidence tricks[15] that draw the mark[16] into parting with their money. R. Paul Wilson wrote and presented the TV series, *The Real Hustle*[17], in which he skilfully shapes and manages the marks expectations. From the BBC website,

> The aim is to reveal how scams work so that the viewer can avoid being ripped off by the same con. The victims of scams in The Real Hustle are known as 'marks' because of an old fairground practice: a chalk mark was put on someone's back if they were deemed to be easy to swindle, marking them out for all the fairground hustlers.
>
> The marks featured in the show have no idea they are being scammed.

If you want to see how easily your expectations can be shaped into believing they should part with their money, then this is a fascinating insight to the world of the confidence trickster.

---

12 https://en.wikipedia.org/wiki/Banachek
13 The author owns number 67 out of the 750 published.
14 https://www.theguardian.com/tv-and-radio/2020/aug/16/derren-brown-20-years-of-mind-control-review-did-he-go-too-far
15 https://en.wikipedia.org/wiki/Confidence_trick
16 https://www.mentalfloss.com/article/72057/11-swindling-slang-terms-grifters
17 https://www.bbc.co.uk/programmes/b006m8mf

In Jim Steinmeyer's *Hiding the Elephant*, he tells the story of the golden age of Magic, of the legendary magicians, for example, John Nevil Maskelyne[18] debunking of the Davenport Brothers, P T Selbit[19] (the first man to saw a woman in half) and his final chapter on the secret of the disappearing donkey, is the reconstruction of one of the world's great illusions.

The main theme that ties all these books[20] together is that they all outline how to influence, manage, and control, the audience's attention and how to communicate a created reality that leads the audience down a carefully constructed belief system.

USING THIS FOR CROWD SAFETY

In our work, the use of these techniques is applied to manage the crowd's expectations, to shape the crowd's behaviour and to enhance the crowds' perception of risk and hence safety. We also apply these techniques when we teach, and use a range of communications methods to both educate, in lectures and articles, and inform in reports and analysis.

However, knowing about the mechanism behind these, and many other, illusions does not make you a magician. Only through practice and performance, can you master even the most basic of illusions. An important quote from *Hiding the Elephant*.

> Leonardo da Vinci was obsessively secret about his techniques for mixing paint. Today this seems silly because we know no formula for pigment or mineral spirits will allow anyone to paint like da Vinci.

For example, the technique called a "retention vanish" uses the visual effect known as "persistence of vision[21]". A retention vanish can take months of diligent practice to perform correctly, to create the illusion of an object disappearing before the spectators' eyes. In magic circles, there is an adage that states,

> The difference between an amateur and a professional is that the amateur will practice until he gets it right, the professional will practice until he doesn't get it wrong.

The magician's perspective presents a fascinating insight into the science of crowd communication, of directing and managing the senses.

We can use the term science as this is a systematic study of the structure and behaviour of the physical and natural world through observation and experiment. These are psychological experiments that are repeatable, with predictable results.

---

[18] https://en.wikipedia.org/wiki/John_Nevil_Maskelyne

[19] https://en.wikipedia.org/wiki/P._T._Selbit

[20] We mentioned only a few, there are of course many volumes written on performance magic and marketing strategies and tactics.

[21] https://en.wikipedia.org/wiki/Persistence_of_vision

SHAPING PERCEPTION

How can we apply these same techniques to enhancing safety in the built and complex environment? As we outlined above, there are many ways to shape individuals and hence the crowd's expectations. From Psychol., 16 September 2016, Editorial: *The Psychology of Magic and the Magic of Psychology*[22],

> Conjurors are masters of illusion and deception, and they have developed astonishing methods for manipulating our experience. Intuitively, the link between magic and psychology seems obvious: magicians use techniques such as misdirection to manipulate our attention, illusions to distort our perception, and forcing to influence our decisions. Some of the early pioneers in Psychology (e.g., Binet, 1894; Triplett, 1900) recognized this close link between magic and psychology and published fascinating scientific papers investigating conjuring techniques. Although some researchers have used magic tricks to study cognition indirectly (e.g., developmental psychologists), few have attempted to bind magic to the science of psychology.
>
> In 2005, Kuhn and Tatler published one of the first recent papers on misdirection, which illustrated how conjuring principles can be used to study visual attention (Kuhn and Tatler, 2005). Whilst this paper attracted much popular interest, many scientists at the time were sceptical about the idea of using magic to explore the inner working of the mind. Although the relationship between magic and psychology is intuitive, this approach requires new paradigms and possibly new ways of thinking about cognitive mechanisms. However, because few researchers have access to the secret armamentarium of magical techniques, studying magic scientifically became the privilege of a small group of investigators with direct experience in conjuring. And yet, the last decade has seen a surge in research papers that have used magic to explore a wide range of topics in psychology. Concrete frameworks now explain how magic can be studied scientifically and the advantages that this direction may provide (Kuhn et al., 2008; Macknik et al., 2008; Demacheva et al., 2012). What was once a field restricted to a few scientists has rapidly grown into a vibrant research domain.

These examples illustrate how information is used to influence crowd behaviour and shape and manage the crowds' expectations.

[22] https://www.frontiersin.org/articles/10.3389/fpsyg.2016.01358/full

MATHEMATICS AND PSYCHOLOGY

We cannot divide the elements of crowd science into *just* the mathematical elements of site analysis, or queueing theory, or the crowd dynamics (the spaces, the flow rates, the crowd density), without an understanding of the motivations and influences on the crowd behaviour. Equally we cannot focus on *just* the psychology of the crowd behaviour without understanding their effect on the crowd dynamics. The two disciplines are intertwined and to understand crowd safety and crowd risk analysis, you need to understand both the mathematics (crowd dynamics) and the behaviour (crowd psychology), hence the term crowd science. From *Misdirection – past, present and future*, by Gustav Kuhn[23],

> What makes the techniques used in misdirection such powerful tools to manipulate spatial attention? Can we identify new attentional principles used by magicians, yet ignored by scientist? How does the context in which the magician is observed influence misdirection? How do magicians control the "collective attention" in an audience? Is this a self-organizing process, alike to what happens when an audience turns into synchronized clapping at the end of a play in a theatre? What are the neural correlates of these synchronizing strategies employed by magicians? The answer to these questions, and many others, may be just a few steps away if we adopt magical techniques, such as those used in misdirection, as part of our laboratory toolkit to investigate sensory awareness.

Psychologists may study how magicians shape the audiences' expectations, how they control attention, and create their illusions. Magicians, on the other hand, keep their secrets to themselves.

OUR RESEARCH OBJECTIVES

Our research aims towards a deeper understanding and practical application of the science of crowd safety and crowd risk analysis. The key questions all relate to how do we influence the crowd behaviour to make public assembly spaces safer? Part of this equates to the position, and information, presented on signage, the distribution of pre-event information via a wide range of sources, the design of spaces, and best practices on managing crowds. It also draws on a wide range of background materials, specifically how marketing companies embed information in our environment, how

23 https://www.researchgate.net/publication/221736041_Misdirection_-_Past_Present_and_the_Future

magicians control attention and how to utilise these techniques and engineer the conditions for crowd safety.

CROWD RISK ANALYSIS

When assessing the safety of crowds in places of public assembly, a comprehensive understanding of crowd dynamics, and crowd behaviour, should be a prerequisite. We know that the process of assessing risks varies from country to country. Across Europe, for example, risk assessments for public gatherings are fundamentally a mechanical and automated exercise, consisting of checklists and "cut and paste" elements from previous events. This often leads to complacency and, although it complies with the most basic requirement of a risk assessment, it has proven to be rudimentary at best. It does not capture the dynamics of the crowd's risks. The risk assessment is an essential tool, not for documenting the crowd risks, but also for managing those risks.

EXPERT WITNESS

As an expert witness, the risk assessment documents are typically the first document we review. These often prove that the event organisers did not understand the risks involved at a site/venue or event and, that lack of understanding, underpins the cause of why incidents occur. Unfortunately, many cases are settled out of court, so the key lessons are never fed back into the system, they are not communicated through the industry. Unlike, say, the aviation industry that investigates all incidents, and close calls (near miss=near collision), using various experts, then feeding the key learning points directly into new processes and procedures. So, all future pilots, or air traffic controllers, can learn to avoid future potential incidents, the event industry does not have those continual improvement processes or procedures. Of course, major incidents are reviewed, and changes are made to the system of crowd safety. For example, the Hillsborough disaster and later Inquiry which led to all seater stadia (Taylor Report).

EDUCATIONAL GOALS

We shall make you a promise that, by the end of this book[24], you *will* understand a suite of crowd risk analysis methods and tools/techniques. Tools that are simple to understand, simple to use and simple to communicate

[24] Or any of our courses, online or in the classroom.

your concerns for the safety of crowds under your charge Whether you are a planner, manager or involved in licensing/permitting a site/venue/event, you will find descriptions of problems, of methods, of solutions and of site applications that you can apply, immediately, and to great effect.

THE SUITE OF TOOLS

The suite consists of five core elements, DIM-ICE, Crowd Dynamics, RAMP Analysis, Risk and Congestion Mapping and the Decision Support (Strategy and Tactics) models.

DIM-ICE: This model was based on extensive research into classifying the causal elements of past crowd disasters. The matrix approach encourages the user to outline the phases and influences on crowd dynamics in both normal, and emergency, situations.

Crowd Dynamics: This is an outline (and visualisation) of the relationships between crowd density, crowd flow and crowd risks. It shows the limits where the onset of shockwaves may occur and how the physical and psychological influences change as density increases in places of public assembly.

RAMP Analysis: Is a tool we developed to focus the user's mind on the routes, areas, movement over time and people/profile of the crowd, in short, the event dynamics. This model was designed to differentiate between the risks (DIM-ICE) and the dynamics (crowds moving towards and away from the site). The user can assess capacity and identify the risk of overcrowding.

Risk and congestion mapping: This is a visualisation of crowd congestion and high-density risks over time using graphics

Decision support analysis: This is a strategy and tactics tool, using matrix analysis to tabulate a range of options and choices.

ORIGINS OF THESE TOOLS

The development of the core modelling tools for crowd safety and risk analysis emerged from problem solving on client projects and research into the underlying causes of crowd major incidents, and from our work as expert witness[25] in international cases. When applying these tools to crowd safety and crowd risk analysis, specifically defining the dynamics of risk in places of public assembly, we see a willing adoption of these models. One of the main reasons for this ready adoption of the crowd risk analysis tools, are

[25] https://crowdrisks.com/expert.html

their simplicity. Over the last thirty years, we have focused on developing simple tools and on teaching and training partitioners, on the use and application of these tools.

THE APPLICATION OF THESE TOOLS

This suite of tools is now mandatory training for all public event commanders in the UK via the college of policing.

> In April 2018 the College of Policing introduced mandatory training in policing events for all Public Order/Public Safety commanders and their advisors. This new training was developed by working closely with Professor Keith Still. ... By April 2019 all strategic, tactical, and operational commanders across England and Wales, and those advising them, will have completed this course.

At the time of this publication, NSW (Australia) are in the process of adopting the UK Crowd Science PEC's course, due to be rolled out in Sept 2021. We are teaching in Police courses in India, (Mid-career Police Training at the Sardar Vallabhbhai Patel National Police Academy), USA, Holland, Belgium, Germany, Ireland, Canada and Australia.

> The Australian Graduate School of Policing and Security, Charles Sturt University, Australia has delivered the Mid-Career Training Programme (Phase III) for over 10 years. As the Co-Director of the programme, aimed at Senior Indian Police Officers and held at the National Police Academy, Hyderabad, India, I needed someone who could teach the higher echelon of the Indian Police Service about the science behind crowd management. Now the Indian Officers know a thing or two about dealing with crowds, or at least they have copious experience of doing so. Of course, experience and expertise are two very different things. Thankfully, several years ago, Professor Keith Still agreed to impart his phenomenal knowledge on dealing with crowds to the course and his presentation has since become a key feature. It is safe to say that Keith is one of, if not the world leader in defining the considerations relating to potential crowded places, planned or otherwise. One can only speculate as to the thousands of lives that have been saved due to Keith's work and not one of those individuals realises. On behalf of all those who have unwittingly survived and will continue to do so in the future because police, planners and governments have taken heed of Keith's work, I express the utmost gratitude for his passion and willingness to share his wisdom.
>
> Graham M Sunderland QPM, MA

The combination of consulting on international projects, of working with both prosecution and defence cases related to personal injury at events, of working with police forces, around the world, and working with front line operators on how to raise the awareness of these methods, may be unique in the industry. We want to share that knowledge, and to encourage others to expand on the use of the crowd safety and risk analysis tools.

As the aim of our work is pass on our understanding of crowd dynamics, and crowd risk analysis. Our objective was to produce and teach a simple set of crowd risks analysis tools, the emphasis of our work is in the production of simple to understand, simple to apply, simply to communicate tools for crowd safety and crowd risk analysis.

FURTHER READING

If you are interested in learning more about the subject of the academic papers on magic, we recommend the following:

Barnhart, A. S. (2010). The exploitation of gestalt principles by magicians. *Perception* 39, 1286–1289. doi: 10.1068/p6766.

Beth, T., and Ekroll, V. (2014). The curious influence of timing on the magical experience evoked by conjuring tricks involving false transfer: decay of amodal object permanence? *Psychol. Res.* 79, 513–522. doi: 10.1007/s00426-014-0584-2.

Binet, A. (1894). Psychology of Prestidigitation. Annual Report of the Board of Regents of the Smithsonian Institution, Washington, DC: U.S. Government Printing Office.

Cavina-Pratesi, C., Kuhn, G., Ietswaart, M., and Milner, A. D. (2011). The magic grasp: motor expertise in deception. *PLoS ONE* 6:e16568. doi: 10.1371/journal.pone.0016568.

Danek, A. H., Fraps, T., von Müeller, A., Grothe, B., and Oellinger, M. (2014). Working Wonders? Investigating insight with magic tricks. *Cognition* 130, 174–185. doi: 10.1016/j.cognition.2013.11.003.

Demacheva, I., Ladouceur, M., Steinberg, E., Pogossova, G., and Raz, A. (2012). The applied cognitive psychology of attention: a step closer to understanding magic tricks. *Appl. Cogn. Psychol.* 26, 541–549. doi: 10.1002/acp.2825.

Diaconis, P., and Graham, R. (2011). *Magical Mathematics: The Mathematical Ideas that Animate Great Magic Tricks*. Princeton, NJ: Princeton University Press.

Ekroll, V., Sayim, B., and Wagemans, J. (2013). Against better knowledge: the magical force of amodal volume completion. *iPerception* 4, 511–515. doi: 10.1068/i0622sas.

Johansson, P., Hall, L., Sikström, S., and Olsson, A. (2005). Failure to detect mismatches between intention and outcome in a simple decision task. *Science* 310, 116–119. doi: 10.1126/science.1111709.

Kuhn, G., Amlani, A. A., and Rensink, R. A. (2008). Towards a science of magic. *Trends Cogn.* Sci. 12, 349–354. doi: 10.1016/j.tics.2008.05.008.

Kuhn, G., and Land, M. F. (2006). There's more to magic than meets the eye. *Curr. Biol.* 16, R950–R951. doi: 10.1016/j.cub.2006.10.012.

Kuhn, G., and Martinez, L. M. (2012). Misdirection - past, present, and the future. *Front. Human Neurosci.* 5. doi: 10.3389/fnhum.2011.00172.

Kuhn, G., and Rensink, R. A. (2016). The Vanishing Ball Illusion: a new perspective on the perception of dynamic events. *Cognition* 148, 64–70. doi: 10.1016/j.cognition.2015.12.003.

Kuhn, G., and Tatler, B. W. (2005). Magic and fixation: now you don't see it, now you do. *Perception* 34, 1155–1161. doi: 10.1068/p3409bn1.

Macknik, S. L., King, M., Randi, J., Robbins, A. Teller., Thompson, J. (2008). Attention and awareness in stage magic: turning tricks into research. *Nat. Rev. Neurosci.* 9, 871–879. doi: 10.1038/nrn2473.

Nisbett, R. E., and Wilson, T. D. (1977). Telling more than we know - Verbal reports on mental processes. *Psychol. Rev.* 84, 231–259. doi: 10.1037/0033-295x.84.3.231.

Olson, J. A., Amlani, A. A., Raz, A., and Rensink, R. A. (2015). Influencing choice without awareness. *Conscious. Cogn.* 37, 225–236. doi: 10.1016/j.concog.2015.01.004.

Olson, J. A., Amlani, A. A., and Rensink, R. A. (2012). Perceptual and cognitive characteristics of common playing cards. *Perception* 41, 268–286. doi: 10.1068/p7175.

Olson, J. A., Landry, M., Appourchaux, K., and Raz, A. (2016). Simulated thought insertion: influencing the sense of agency using deception and magic. *Conscious. Cogn.* 43, 11–26. doi: 10.1016/j.concog.2016.04.010.

Parris, B. A., Kuhn, G., Mizon, G. A., Benattayallah, A., and Hodgson, T. L. (2009). Imaging the impossible: an fMRI study of impossible causal relationships in magic tricks. *Neuroimage* 45, 1033–1039. doi: 10.1016/j.neuroimage.2008.12.036.

Shalom, D. E., de Sousa Serro, M. G., Giaconia, M., Martinez, L. M., Rieznik, A., and Sigman, M. (2013). Choosing in Freedom or forced to choose? Introspective blindness to psychological forcing in stage-magic. *PLoS ONE* 8:e58254. doi: 10.1371/journal.pone.0058254.

Subbotsky, E. (2010). *Magic and the Mind.* Oxford: University Press.

Teller (2012). Teller reveals his secrets. *Smithsonian Magazine.*

Thomas, C. C., Didierjean, A., Maquestiaux, F., and Gygax, P. (2015). Does magic offer a cryptozoology ground for psychology? *Rev. Gen. Psychol.* 19, 117–128. doi: 10.1037/gpr0000041.

Triplett, N. (1900). The psychology of conjuring deceptions. *Am. J. Psychol.* 11, 439–510.

FROM WORKSHOP TO MSC

We asked one of our contributors and an MSc graduate from our Crowd Safety and Risk Analysis course, to outline her journey from industry to academia and onwards.

ÍSE MURPHY[26]

Working in transport planning for major events, I often spent my time designing queuing systems, calculating ingress and egress capacities as crowds move from station to stadium and back. I was responsible for the safe movement of thousands of people as they arrived at a city or event footprint, ticket in hand. In the back of my mind, I knew that I should have some form of qualification and professional training to be undertaking my role, as I was dealing with crowd dynamics and crowd psychology. When I accidentally happened upon Keith Still's 3-day course in January 2016 in a conversation over lunch with a friend, I signed up, with the course starting the next day.

Meeting Keith, I knew I had much to learn, and exactly what I needed to learn to be able to carry out my role effectively. What started as a three-day course, grew into a qualification on submission of coursework, which then qualified me to join the MSc Crowd Safety and Risk Analysis. I took my seat in Manchester Metropolitan University in September 2017 along with colleagues who have grown into friends and begun the transformative journey. I have always worked in the events industry, an industry I love. I studied BSc Events Management back in 2007 and went on to work on major events and festivals across the world. My roles often varied, and it wasn't until meeting Keith that I discovered my passion for Crowd Safety. My niche was born and since completing my MSc with Distinction, I work full time as Senior Crowd Safety Manager at Wembley National Stadium, guest lecture for universities across the UK and Ireland and am Associate Lecture for Crowd Management at the University of Plymouth. I have travelled as far as Australia and Saudi Arabia, working with event organisers to develop crowd safety practices.

From Crowd Science to Crowd Safety, working with Keith has opened a whole new world, a world I am passionate about, and a world I aim to make safer for all those who attend events.

Íse Murphy
June 2021

[26] https://www.isemurphy.com

Chapter 1

Virtual learning environment (VLE)

> Knowledge is like money: to be of value it must circulate, and in circulating it can increase in quantity and, hopefully, in value.
>
> Louis L'Amour

If you bought this book as a companion to the online learning course, this chapter is a guide to help you through the first logging-in and the learning process. However, if you want to read about the background, the underlying theories, the essays, and case studies, you can skip this chapter.

On our VLE (GKSed.com), we host a Moodle-based series of crowd safety courses. These cover Level 3 and Level 5 qualifications. The crowd science courses are accredited by both the International Institute of Risk and Safety Management[1] and Highfield.[2] There were over 1,000 learners through the system in 2020 and in 2021, NCS4 (University of Southern Mississippi[3]) partnered with us to launch the USA version of the Level 3 course.

USING THE ONLINE LEARNING PROGRAMME

If you have not used an eLearning system before, it is straightforward. Our courses follow the format and structure of the content taught in classroom courses. There are, of course, clear advantages and disadvantages to online learning.

BASED ON FEEDBACK

As the main feedback from our classroom courses was "too much content, not enough time to study each topic", we developed the online version partly for learners who could not attend a classroom session, but also to aid

1 https://www.iirsm.org
2 https://www.highfieldqualifications.com
3 https://ncs4.usm.edu

DOI: 10.1201/9781351053068-1

those who attended the course and wanted to recap and study each topic in depth. Everyone who attends our classroom courses has access to the online version to refer to key learning points, at their own pace.

The significant advantage of online learning is that you can take your time to study the materials and replay any section that has information that might be new to you. In the classroom, it is often difficult to repeat, or expand on a topic, due to time constraints and the presence of other learners. Solo study allows you to learn at your own pace.

To aid the learner, we have forum discussions, question and answers sections, and background reading materials for each topic to compensate for this. If you search back on forum topics or ask a new question, we can post the reply, and everyone receives help from that discussion. However, the first step is to get you online.

You can buy any of our courses via the website (www.crowdrisks.com) or NCS4 in the USA. Once bought, the system <noreply@gksed.com> will send you your log-in credentials, along with an email and frequently asked questions. Your username is your email address (all in lowercase), and your password is a string of random symbols sent on the registration email.

Use those two bits of information to log into GKSed.com for the first time. It will prompt you to change your password (which is a regular prompt, every few weeks for obvious security reasons).

INITIAL REGISTRATION

Once you log in and change your password, which is the first screen you will see. You will find that the courses that you have bought as the blue sections. These are the hyperlinked. So, if you click on (say) the *Introduction to Crowd Science* block it will take you into the course area. You then have a series of options where you can scroll down (on the left-hand side) the various course elements and topics (Figure 1.1).

In each of the sub-sections, and each of the modules you will find the content. For example, "Lectures" is the area where you can view the lecture videos. The menu and content are structured, so best start at the beginning and work your way down, and through the content.

You can download the PDF's[4] of the lecture slides, support documents, either by topic or all at once. For example, in the topic associated with risk analysis, you will find various documents on risk analysis. In the folder on site design, you will find the documents on site design. We update the documents with the latest information, new case studies and when new papers are published.

There are sub-sections for forum discussions, where you can post information about any specific topic. The little blue and green bubbles speech

[4] Portable document format – an adobe format which you can view/read on your own computer or print out for reference.

Course information

Welcome to the on-line version of the "Introduction to Crowd Safety and Risk Analysis" course. The course uses the Moodle platform and is self-paced. If you are new to Moodle, take a few minutes to view the "how to-" section (in **lectures**) and read the discussion forums.

You can also introduce yourself to others learners interesting in Crowd Science in the introduction section. We have many international learners, and its a great way to network with like-minded colleagues.

To pass the course, you need to submit the two qualifying essays for assessment and certification. You can submit a draft for comment (we are always happy to help - but I can't write it for you). The essays have a specific format and you'll find examples in the relevant section.

Feel free to post any questions on any of the topics forums we will reply promptly.

You will find all the associated papers for each section on this link

Announcements

Please introduce yourself here

1. Course Introduction

In this section we introduce the course format, outline the course topics and show you how to proceed through the Moodle system and access the course materials. Click on the **lectures** and slides link below (or on the panel on the left) to take you to the video clips of the course materials for each topic. In the resources folders, you will find further reading materials.

Figure 1.1 Introduction screen on the VLE.

bubbles allow you to create a discussion topic, ask a question, post information that fellow students can share, post documents, and request details for information. We will get back to you within a couple of hours of you posting questions because these go straight to the moderators of the various forums.

In the lecture's sections, you will find a playlist. You play all the videos on that list or stop and start when you want. This works on mobile devices, on laptops, and desktops. You can use the sidebar to navigate through the learning programme.

There is a search facility to allow you to access the various forums. There are various menu options on the left-hand side of the screen to allow you to skip to a topic. As we cover each topic in depth, (more than we can deliver in a classroom environment) you can ask any questions on any course topic, post those to the forum, use the forum discussion, see previously asked questions, and take part in the conversations.

ASSESSMENTS AND CERTIFICATES

The level 3 course, Fundamentals of Crowd Safety, has a 60-day access. To pass this course, there is a timed (1 hour) multiple choice quiz to pass. There is a bank of random questions, in a random order, which relate to the course materials, and you need to achieve more than 80% to pass. You have three attempts, and you need to click on the "Check" button to register your answer. Some questions have a weighting (a scale of correct) and some need multiple answers. The system produces a certificate that you can download. Each certificate has a unique code, which future employers can cross check your qualifications. We also send each pass to the awarding body, for your Level 3 certificate. We email those to you and again, they have a unique number for cross checking should you lose the original.

The level 5 course, Introduction to Crowd Science, the certificate, is based on two written assessments. The first is an essay on the general discussion on the crowd sciences, and the second is the case study, where you need to show the use of the tools in the course. Examples of both of those essays are in later chapters of this book.

Plagiarism and copying other learner's work will result in failure. However, cross-referencing, and quoting other sources, proves a depth of reading and understanding. You have 6 months to complete the Level 5, Introduction to Crowd Science course. You can send a draft version for review and initial feedback, but only once or twice. Once you send your final essays, we will assess them, then supply feedback and certification. Again, we send these to the awarding body for your Level 5 certificate.

We are always happy to answer questions and expand on any of the key topics for discussion. As with any learning experience, you will get the most benefit by asking questions.

Chapter 2

The DIM-ICE risk model

> The knowledge of anything, since all things have causes, is not acquired or complete unless it is known by its causes.
>
> Avicenna

The principles of crowd safety and crowd risk analysis are independent of scale. A small crowd needs the same risk analysis as a larger crowd. The physical force (pressure) needed to inflict serious personal injury is small.[1] Crowd risk analysis is the process of showing the risks of personal injury in places of public assembly. Whether the crowd is small or large, the principles of crowd safety and crowd risk analysis are the same across all scales. The hazards and risks that can cause harm are not simply a function of the number of people present in any given environment. Crowd risks are a function of localised crowd density, of forces that localised high density can exert, and of the time of the exposure to those forces. This is not *just* the number of people present in any given event. In other words, crowd safety risks manifest when localised crowd *density* exceeds safe limits. It could be a crowd of four hundred young adults (Cookstown, 2019) or tens of thousands of pilgrims (Hajj, 2001, 2004, 2006, 2015).

The most common question of "how many people can we have at this event?" is often asked both in class and online. It is not the number of people; it is the number of people per square metre/yard, the localised high density. It is not the total space you may have available; it is how people occupy and move around and gather within that space that can give rise to crowd risks. You need to understand how to apply a risk assessment and *how* to plan and manage your crowds based on safety, on crowd flow and localised high density.

Obviously, the *consequences* of an incident can be much greater with a larger crowd, but the nature and the physics of the risks are similar; the *principles* of crowd safety and crowd risk analysis are independent of scale.

[1] Kroll MW, Still GK, Neuman TS, Graham MA, Griffin LV. Acute forces required for fatal compression asphyxia: A biomechanical model and historical comparisons. *Medicine, Science and the Law*. 2017;57(2):61–68. doi:10.1177/0025802417695711.

DOI: 10.1201/9781351053068-2

We have worked on consulting projects around the world, in major accident and incident investigations, as an expert witness for dozens of international cases involving crowds. We teach the principles and applications of crowd safety and crowd risk analysis based on that range of experiences. Crowd risk analysis is not an academic exercise or series of theoretical models. When we worked in Saudi Arabia on Hajj projects, we must accumulate and classify information from a wide range of various sources. When we worked on accident and incident investigations, we need to assess the causality of the incident and outline this, in laypersons terms to the legal teams and the court. When we design event spaces, we need to apply a model that works or be liable for any consequential injuries. When we teach a class of practitioners, we need to back our modelling tools and techniques with evidence[2] of their successful application.

LEGAL REQUIREMENTS

Understanding the legal process (UK, Europe, USA, Australia) and how the court systems work shape the focus of our courses. Obviously, we cannot talk about details of any specific cases, but the underlying causes, the similarities between situations that give rise to incidents and injuries, help us shape future crowd risk analysis tools and key learning points. For example, understanding the different types of crowd risks, crowd dynamics, crowd behaviour, how incidents occur, and what the learners need to do to understand these risks and apply appropriate crowd risk mitigation and management strategies are the essence of our international courses.

We focus on the underlying causes of crowd safety problems, how to prepare your safety plans, how to prevent accidents and incidents at your site, and how to develop your crowd management plan. This all leads to safer crowd management at your events and public gatherings. We do not claim that this is a substitute for field experience, but we do bring a rigorous method that makes our learners better crowd managers, adding a scientific rigour to their experience.

Two quotes come to mind when we hear people talking about experience.

> Experience enables you to recognise a mistake when you make it again
>
> *Franklin P. Jones*

> Judgement comes from experience, and experience comes from bad judgement
>
> *Simon Bolivar*

[2] Still P. Place crowd safety, crowd science? Case studies and application. *Journal of Place Management and Development Emerald Publishing Limited* 1753–8335 DOI https://www.emerald.com/insight/content/doi/10.1108/JPMD-10-2019-0090/full/html

There may be no classroom substitute for the experience of managing crowds, but when you are managing crowds, you cannot afford to make mistakes. Not understanding crowd safety and crowd risks can often result in dire consequences, personal injuries, and loss of life.[3] Hence there is a need to add a scientific basis to your experiential learning.

ACCREDITED PRIOR LEARNING

Crowd science courses also bridge the gap between a specific experiential learning and a broader application of crowd risk analysis of different events, different crowds and unusual places. For example, the minor course at Breda University of Applied Science couples event experience and the theoretical/scientific courses for assessing crowd safety.

If you want to move on to higher education, we recommend the experience of events and the Introduction to Crowd Science course. These courses are useful to bridge the gap between the practitioner and academia.

PROBLEMS "LYING IN WAIT"

Crowd risk analysis is the ability to find crowd risks that are lying in wait and manage or mitigate those risks *before* you expose your crowd to those risks. We combine the historical analysis of crowd-related incidents and a range of risk assessment methods to assess these risks and communicate the nature of those risks to multiagency teams. The suite of tools outlined in this book lets you analyse your event/site design and document your due diligence for your safety plan. The case studies illustrate how to use these tools for event planning around the world.

CAUSALITY

One of the themes that runs through our sources is understanding the causes of major incidents. In a legal context that are complex definitions of causality. From Thomson Reuters Practical Law[4]

> CAUSATION
>
> A principle used in the assessment of damages for breach of contract or tort. Losses may have been foreseeable at the time of contracting or

[3] From a legal perspective defined as "was there a significant risk to life and limb" at this location/event/venue?

[4] https://uk.practicallaw.thomsonreuters.com/4-107-5865?transitionType=Default&contextData=(sc.Default)&firstPage=true

at the time of the breach of duty in the case of tort, but they will only be recoverable if those losses were caused by the breach of contract or duty.

And, paraphrased from the USA Legal Dictionary[5]

Cause

... Something that precedes and brings about an effect or a result. A reason for an action or condition. A ground of a legal action. An agent that brings something about. That which in some manner is accountable for a condition that brings about an effect or that produces a cause for the resultant action or state

There are fundamental differences between definitions of cause and causality. From the USA Legal Dictionary[6]

Cause and Causality in American Law

In the law of torts, the concept of causality is essential to a person's ability to successfully bring an action for injury against another person. The injured party must establish that the other person brought about the alleged harm. A defendant's liability is contingent upon the connection between his or her conduct and the injury to the plaintiff. The plaintiff must prove that his or her injury would not have occurred but for the defendant's Negligence or intentional conduct.

We also highlight that a chain of events leading to injury is defined as:

Actual, Concurrent, and Intervening Cause

The actual cause is the event directly responsible for an injury. If one person shoves another, thereby knocking the other person out an open window and he or she breaks a leg as a result of the fall, the shove is the actual cause of the injury. The immediate cause of the injury in this case would be the fall, since it is the cause that came right before the injury, with no intermediate causes.

There are definitions of Proximate, Unforeseeable and Remote causes. From USA Legal Dictionary[7]

Proximate, Unforeseeable and Remote Cause

The proximate cause of an injury is the act or omission of an act without which the harm would not have occurred. ... An unforeseeable cause

---

5 https://legal-dictionary.thefreedictionary.com/Cause+and+Causality+in+American+Law
6 https://legal-dictionary.thefreedictionary.com/Cause+and+Causality+in+American+Law
7 https://legal-dictionary.thefreedictionary.com/Cause+and+Causality+in+American+Law

> is one that unexpectedly and unpredictably results from the proximate cause. ... A remote cause is one that is removed or separate from the proximate cause of an injury.

TWO KEY DEFINITIONS

The reason for the above citations is that the two key definitions (proximate and distal) relate to the cause of injury and the test of foreseeability.

> The proximate cause of an injury is the act or omission of an act without which the harm would not have occurred.
>
> An unforeseeable cause is one that unexpectedly and unpredictably results from the proximate cause.

If a risk is foreseeable, by an expert, a trained crowd manager or by a layperson, then the consequences are predictable. We have seen the defence that "no-one could have predicted this or that consequence" and demonstrated that there are training programmes, risk[8] assessment methodology, guidance, and legislation that exist because the incident was, in fact, reasonably foreseeable, that the consequences were foreseeable.

We do not aim to challenge or reframe the complex legal definitions of causality. We only use these terms to outline the two basic *principles* of causality. The proximate cause, the sequence of events that occur immediately before the incident, and the distal causes, the underlying causes. For example, inadequate processes and procedures, ignorance of the risks or negligent practices, the "*we have always done it this way*" attitude lay the foundation of failure.

Our work focuses on the development of crowd risk analysis tools with an emphasis on naming those problems that are "lying in wait", the accidents waiting to happen. So how do we recognise problems lying in wait? How do we explain the principle and differences between proximate and distal causality?

THE TITANIC

We use an example that we once heard Gordon Graham[9] use, the sinking of the Titanic. When the Titanic sank, three and a half thousand people lost their life. We would ask the class to outline what they think caused the loss of life? Who was to blame? Of course, proximate causes are the

[8] Far From the Madding Crowd: A Statutory Response to Crowd Crush, 68 Hastings L.J. 159 (2016). and Crowd Crush: How the Law Leaves American Crowds Unprotected, 104 Kentucky L.J.1 (2016).

[9] http://www.gordongraham.com/about.html

first thing the class typically outlines. The list includes the following: hitting an iceberg, late call to "abandon ship", management indecision, poor decision making, not enough lifeboats, the water was too cold...typically all in the definition of a proximate cause. The events that directly cause a loss of life. The things that were going wrong before and as the ship was sinking.

THE UNDERLYING CAUSE

There was a fundamental underlying (hence distal) cause. Fundamentally, the failure was the safety culture. Namely, the Titanic was the unsinkable ship. Everything, all the process procedures, management, emergency planning, lifeboats, navigation, the attempt at the fastest Atlantic crossing, the route they chose, all based on the assumption that this was an unsinkable ship.

With the underpinning concept for the design was that the Titanic was an unsinkable ship. They did not need to worry about icebergs because it was an unsinkable ship. All decisions, the design, the information systems, the management process and procedures were all built on the assumption that the ship was unsinkable.

To quote Captain EJ Smith (1907):

> When anyone asks me how I can best describe my experiences of nearly forty years at sea I merely say uneventful. Of course, there have been Winter gales and storms and fog and the like, but in all my experience I have never been in an accident of any sort worth speaking about. I have seen but one vessel in distress in all my years at sea, a brig, the crew of which was taken off in a small boat in charge of my third officer. I never saw a wreck and have never been wrecked, nor was I ever in any predicament that threatened to end in disaster of any sort.
>
> I will say that I cannot imagine any condition which could cause a ship to founder. I cannot conceive of any vital disaster happening to this vessel. Modern shipbuilding has gone beyond that.
>
> *Captain EJ Smith was the Captain of the Titanic, the Titanic sank in April 1912*

In our courses, we expand on the sinking of the Titanic and outline that there were design failures (hull breach), there were information failures (not recognising the physics of icebergs, late call to "abandon ship"), there were management failures (route choices, poor decision making during the emergency), etc., and then draw on the history of crowd related accidents and incidents to expand on this analogy.

Design failures, lack of information and management failures are distal causes of crowd-related incidents.

ANYTHING THAT CAN GO WRONG WILL GO WRONG – MURPHY'S LAW

The following is a series of quotes[10] with reference to Murphy's law which serve to illustrate the basic thesis of our work in crowd risk analysis.

> If there's more than one way to do a job and one of those ways will end in disaster, then somebody will do it that way.
>
> *Version attributed to Edward A. Murphy in People Magazine (31 January 1983), p. 82*

> Colonel Stapp's favourite take-off on sober scientific laws — Murphy's Law, Stapp calls it — 'Everything that can possibly go wrong will go wrong' Lloyd Mallan, in Men, Rockets and Space Rats (1955), possibly the earliest printed use of Murphy's name in connection with the law.

Murphy's law illustrates the philosophy behind risk analysis, namely, anticipate the how, the when, the where, and the why things may fail, then plan risk mitigation to minimise the risk of failure.

There is a difference between a failsafe system (one designed not to fail) and a safe failure. A "safe fail" is a system you expect may fail and design elements that will allow it to fail safely. In crowd risk analysis, the difference between "fail-safe" and "safe fail" comes down to planning. As an analogy, we could try and design a system that does not reach high pressure, but that may prove to be much more challenging than designing a system in which we are expecting that pressure might build up, and then design a safety valve to release the pressure automatically. That is a safe failure approach.

An example of this, from the Wembley Stadium complex, was a series of push bar exit doors built into an underground passage, leading towards a narrow station entrance. If there was a build-up of crowd pressure, the mass of bodies would automatically push the bars, the doors would open and the crowd pressure released.

DENSITY AND FLOW

When considering a design, we look to see if it is safe for that number of people trying to get through that entry/exit at the expected time. Or if this area is suitable for the expected number of people at this location. Whether it is that number of people trying to get on a lifeboat or that number of people trying to move into Hillsborough, out of Ibrox, or in and out of the ramp at the Love Parade, we see that any design has physical limits. If we try to exceed those

[10] https://en.wikiquote.org/wiki/Murphy%27s_law

physical limits, if the demand (number of people) exceeds the systems volume/capacity, then clearly, there will be a problem. Design limits are calculable using crowd density and crowd flow, area, distance, and width calculations.

Design can influence crowd behaviour and shape the crowds' expectations. Information can motivate a crowd to move in a particular way or behave in a particular fashion. How you manage a crowd can influence both individual and group behaviour. Design, information and management are the three underlying causes of crowd-related accidents and incidents. Poor design, misinformation and inappropriate crowd management can create issues and problems with the crowds' actions and reactions.

When designing risk assessment tools, we need to focus the user's attention on the issues of design, information and crowd management. We can look at events through the filter of crowd influences (design, information and management) and visualise potential problems.

Why is it important? Well, if you understand the distal causes of accidents or disasters, you can improve safety through your own risk management process. If you can understand the liabilities, you can invest resources to prevent accidents at your event/site. Lord Justice Taylor said at the end of the Hillsborough disaster report[11] (1989) said the following:

> CHAPTER 2 (PAGE 24) – A BETTER FUTURE FOR FOOTBALL
>
> 59. It is not enough to aim only at the minimum measures necessary for safety. That has been, at best, the approach in the past and too often not even that standard has been achieved. What is required is the vision and imagination to achieve a new ethos ...

1989 was the year we started our research into crowd dynamics, with the question, what is this new vision and imagination? What was missing in our understanding of crowd accidents and incidents, which needed a new vision?

To understand that question, to understand what might be missing from the existing guidance and legislation, we researched reports of major incidents, from around the world. Our research data mined the numerous reports of major crowd related accidents and incidents, looking at the underlying causes, and we began to define a classification system, the fundamental causes of accidents and incidents.

OPERATIONS RESEARCH

With a background in operations research, we used techniques we had used in manufacturing process re-engineering. At that time, we were working on a Virtual Reality project with Eddie Obeng.[12]

[11] Inquiry by the Rt Hon Lord Justice Taylor. The Hillsborough Stadium Disaster 15 April 1989, Final Report Cm 962. IMSO London.

[12] https://en.wikipedia.org/wiki/Eddie_Obeng

> Obeng pioneered the use of bespoke business simulation games to help stimulate and embed learning. In 1995, together with Keith Still, pioneer in crowd dynamics, he created the now obsolete virtual reality game 'Columbus'. Programming required high powered SGI Indigo computers, although the simulation ran at 25 fps on a standard PC running Windows 95. In 2010, he launched 'The Cube' (later rebranded to QUBE). Obeng's innovation with QUBE was to focus on learning and application by integrating all the business models, frameworks, and tools into the virtual reality enabled environment.

The "Pizza Project", as we called it back in the early 1990s, used the same modelling techniques that we used for crowd modelling. Developing agents that reacted to their environment, and interacted with each other, in a virtual reality environment using both the principles of business process re-engineering and teaching/training techniques.

> Eddie is an inspirational thinker and author. The overlap between his classification of manufacturing resource planning, the movement of materials, information, and resources, to optimise the cost function, and our games development, used the same mathematical frameworks. For example, if your rough-cut capacity planning and bottleneck analysis identified when your demand might exceed your capacity, something needs to change, or you will lose the business opportunity. Hence, your business process would need to be re-engineered, and the change managed to implement this in your organisation. We developed an environment to demonstrate the integration of the various business processes and how these impact each other. We were teaching the user how to understand business complexity through an abstraction.

HOW DOES THIS APPLY TO CROWD RISK ANALYSIS?

Working with Eddie had parallels to the work we were doing on the classification of causality of crowd disasters. Where Eddie had found three core elements to all business processes, with the aim of re-engineering, optimising, and improving operations, we could use a similar framework for crowd safety.

However, the problem was not just to classify the causes of major incidents and accidents but to develop a novel approach to teach these principles to the industry. As we later discovered, this turned out to be much more than a new vision and imagination. It required a new ethos. There was nothing to be gained in designing complex methods to assess risk; we needed a simpler method of communicating the dynamics of crowd risks.

As Martin Luther King Jr once said:

> Knowledge is a process of piling up facts; wisdom lies in their simplification.

In effect, there was no point in creating a classification framework, creating complex mathematical models of risk, or coding those into computer simulations that only an expert could use or understand. We needed to *simplify* the process of the risk analysis and risk assessment. In the same way, we taught the users of the virtual reality game "Columbus", we asked, could we teach the principles of crowd safety and crowd risk analysis to crowd managers? The process had to both engage the attention of the user and focus their perception of crowd risks in the built and complex environment. It could not be another set of tables and degrade into another box ticking exercise.

CLASSIFICATION OF MAJOR INCIDENTS

When we examine past crowd disasters, review the underlying causes, we could see where we need to plug the knowledge gaps and improve the risk analysis process. To start, we needed to classify major incidents. We needed to develop a framework, to see if there were common features, things that we could learn from *why* a system had failed. Then we needed to figure out how to encode those in easy-to-use risk analysis tools and techniques. We needed to develop a method of educating and informing those responsible for crowd safety, to help a knowledge transfer.[13] This was neither quick, nor easy, with over thirty years of research, consulting, advising and teaching, there are still major incidents. We still have an audience to inform. We still see and hear about sites/venues and events that are at risk.

A HISTORY OF INCIDENTS

Table 2.1 is a summary[14] of incidents from 1989 through to the date of writing this chapter, with the classification of underlying, distal causality.

The majority of the incidents outlined in Table 2.1 have underlying design related causes, namely, too many people and not enough space (overcrowding/overcapacity). Table 2.1 is not a complete list of disasters. We only reviewed those incidents that somebody could have prevented by using an appropriate risk assessment and acting upon the analysis before the crowds arrived. Sadly, this list continues to grow.

We have over one hundred incidents in our case study materials for the online teaching programmes. In the MSc programme, we provide the complete database to the students, and challenged them to assess the DIM-ICE classification, and to add/enhance the risk models we defined.

[13] Hence our courses, and this book.

[14] Note: this is not the complete list of disasters; only a summary for the purposes of classification.

Table 2.1 Summary of the list of disasters since 1989

Year	*Location*	*Scale of the incident*	*Failure elements*
1989	Hillsborough, UK	96 dead, 400 injured (Overcrowding)	Design (Throughput)
1990	Mina Valley, Saudi Arabia	1,426 pilgrims crushed (Overcrowding)	Design (Throughput)
1991	Orkney, South Africa	42 dead, many injured (Overcrowding/riot)	Design (Capacity + Crazing)
1994	Jamarat, Saudi Arabia	266 pilgrims crushed, 98 injured (Overcrowding)	Design (Throughput)
1996	Guatemala City	83 crushed, 180 injured (Overcrowding)	Design (Capacity + Forged Tickets)
1997	Jamarat, Saudi Arabia	22 pilgrims crushed, 43 injured (Overcrowding)	Design (Throughput)
1998	Jamarat, Saudi Arabia	118 pilgrims crushed, 434 injured (Overcrowding)	Design (Throughput)
1999	Kerala, India	51 killed, 150 injured in stampede (Reaction)	Information (Weather + Running)
1999	Minsk, Belarus	53 killed, 190 injured in stampede (Reaction)	Information (Weather + Running)
2001	Ellis Park, South Africa	43 dead, 200+ Injured (Over capacity - Overcrowding)	Design (Capacity + Crazing)
2001	Aracaju, Brazil	4 dead, including 3 children (Reaction)	Information (Handouts + Crazing)
2002	Yokohama, Japan	10 trampled, Mall crowd craze (Reaction)	Information (Handouts + Crazing)
2001	Jamarat, Saudi Arabia	35 pilgrims crushed, 179 injured (Overcrowding)	Design (Throughput)
2004	Jamarat, Saudi Arabia	249 pilgrims crushed, 252 injured (Overcrowding)	Design (Throughput)
2004	Beijing, China	37 dead, 15 injured in crowd crush (Overcrowding)	Design (Throughput)
2006	Jamarat, Saudi Arabia	363 dead, 389 injured crowd crush (Overcrowding)	Design (Throughput)
2006	Manila, Philippines	74 dead, 300 injured crowd crush (Reaction)	Design (Throughput)
2006	Yemen, Middle East	51 dead, 238 Injured crowd crush (Reaction)	Information (Political Rally)
2008	Himachal Pradesh, India	146 dead, 50 injured in stampede (Narrow Road)	Design (Throughput)
2008	Pasuran, Java	23 dead, dozens injured in Ramadan (Reaction)	Information (Handouts)
2009	Abidjan, Ivory Coast	22 dead, 132 injured Football (Reaction)	Design (Throughput)
2009	Birmingham (JLS), UK	60 injured, 4 hospitalised (JLS) (Reaction)	Design (Capacity +Arrival Profile)

(*Continued*)

Table 2.1 (Continued) Summary of the list of disasters since 1989

Year	*Location*	*Scale of the incident*	*Failure elements*
2010	Timbuktu, Mali, West Africa	26 dead, 55 injured (Overcrowding)	Design (Throughput)
2010	Kunda, North India	63 dead, 44 injured (Overcrowding-narrow Road)	Design (Throughput)
2010	Johannesburg, South Africa	14 injured (Overcrowding at entry gates)	Information (Forged Tickets)
2010	Duisburg, Germany.	21 dead, 511 injured (Overcrowding)	Design (Throughput)
2010	Bihar, India	10 dead, dozens injured (Reaction)	Information (Over Reaction)
2010	Nairobi, Kenya	7 dead, 70 injured (Reaction)	Information (Rain Stopped)
2010	Phnom Penh, Cambodia	347 dead, 395 injured (Overcrowding)	Design (Capacity)
2011	Kerala, India	102 dead, 44 injured (Overcrowding)	Design (Capacity)
2011	Bamako, Mali	36 dead, 70 Injured (Overcrowding)	Design (Capacity)
2011	Brazzaville, Congo	7 dead, 30 Injured (Overcrowding)	Design (Capacity)
2011	Jakarta, Indonesia	2 dead, 13 Injured (Overcrowding)	Information (Poor Ticket Allocation)
2012	Port Said, Egypt	74 dead, over 1,000 Injured (Overcrowding + rioting)	Design (Throughput + Riot)
2012	Cairo (copic), Egypt	3 dead, dozens injured (Overcrowding)	Design (Throughput)
2013	Abidjan, Ivory Coast	62 dead, dozens injured (Overcrowding)	Design (Capacity + Crazing)
2013	Allahabad, Northern India	36 dead, 31 injured (Overcrowding, railway platform)	Design (Capacity + Crowd management)
2013	Hubei, China	4 Dead, 14 injured (Overcrowding on stairs - school)	Crowd Management
2013	Datia, India	50 dead, 100+ injured (Overcrowding)	Design (Capacity + Crowd management)
2013	Anambra, Nigeria	28 dead, 200+ injured (Overreaction - call of "fire")	Design (Capacity and crazing)
2014	Ningxia, China	14 Dead, 10 Injured (Overcrowding – food handouts)	Design (Crowd flow, Crowd management)

(Continued)

Table 2.1 (Continued) Summary of the list of disasters since 1989

Year	*Location*	*Scale of the incident*	*Failure elements*
2014	Mumbai, India	20 dead, 40 injured (Overcrowding, narrow streets)	Design (Capacity, Crowd Flow)
2015	Shanghai, China	36 dead, 46 injured – crowd crushing (overcrowding)	Design (Capacity, Crowd Flow)
2015	Mina Valley, Saudi Arabia	1,500+ dead, thousands injured (overcrowding)	Design (Capacity, Crowd Flow)
2016	Lucknow, India	24 dead, 20 injured (overcrowding on bridge)	Design (Capacity, Crowd Flow)
2016	Falls Festival, Australia	60 injured on egress (slope, scheduling, overcapacity)	Design (Capacity, Crowd Flow)
2017	Uíge, Angola	17 dead, 61 injured (ingress, overcapacity)	Design (Capacity, Crowd Flow)
2017	Lusaka, Zambia	8 dead, 28 injured (food aid, queueing)	Design (Capacity, Crowd Flow)
2017	Tegucigalpa, Honduras	4 dead, 25 injured (ingress, overcapacity)	Design (Capacity, Crowd Flow)
2017	Turin, Italy	2 dead, 1,527 injured (firework mistaken for bomb)	False alarm (Information)
2017	Lilongwe, Malawi	8 dead, 40 injured (ingress, overcapacity)	Design (Capacity, Crowd Flow)
2017	Dakar, Senegal	8 dead, 40 injured (tear gas, management)	Management (police - tear gas)
2017	Mbour, Senegal	8 dead, 60 injured (ingress, overcapacity)	Design (Capacity, Crowd Flow)
2017	Johannesburg, South Africa	2 dead, 17 injured (ingress, overcapacity)	Design (Capacity, Crowd Flow)
2017	Mumbai, India	22 dead, 39 injured (overcapacity, bridge)	Overcrowding, Information
2017	Sidi Boulaalam, Morocco	15 dead, 40 injured (queueing)	Design (Capacity, Crowd Flow)
2018	Caracas, Venezuela	17 dead, dozens injured (tear gas – management)	Management (Tear gas)
2018	Nice, France	30 injured (firework mistaken for bomb)	False alarm (Information)
2018	Celtic Park, Scotland	5 injured (design, management)	Management (Crowd Flow/design)
2018	Madagascar	1 dead, 40 injured (Design, Management)	Design (ingress crush)
2018	Luanda, Angola	5 dead, 7 injured	Management (gates shut)
2018	Corinaldo, Italy	6 dead, 59 injured	Design (bridge/barrier collapse)
2019	Port Harcourt, Nigeria	15 dead, 12 injured	Design (locked gate)

(Continued)

Table 2.1 (Continued) Summary of the list of disasters since 1989

Year	*Location*	*Scale of the incident*	*Failure elements*
2019	Kuala Lumpur, Malaysia	2 dead (elderly)	Design (queueing system)
2019	Cookstown, Ireland	3 dead, dozens injured	Design (queueing system)
2019	Antananarivo, Madagascar	16 dead, 101 injured (Design, Management)	Design (Doors closed)
2019	Algiers, Algeria	5 dead, 86 injured	Design (4 small entry points)
2019	Caracas, Venezuela	4 dead, 50 injured	Design and Management (ingress)
2020	Kerman, Iran	56 dead, 200 injured	Design and Management (over capacity)
2020	Kakamega, Kenya	14 dead, 40 injured	Design/Management (overcrowding on stairs)
2020	Diffa, Niger	20 dead, 10 injured	Information, management, handouts
2020	Lima, Peru	13 dead, 3 injured	Overcrowding, illegal nightclub, closed exits
2020	Maligawatta, Sri Lanka	3 dead, 9 injured	Information, management, handouts
2020	Moshi, Tanzania	20 dead, 16 injured (anointing oil, handout)	Information, management, handouts
2021	El Alto, Bolivia	7 dead, 5 injured	Barrier collapsed, overcrowding
2021	Dar es Salaam, Tanzania	43 dead, 37 injured	Design and Management (over capacity)
2021	Mount Meron, Israel	44 dead, over 100 injured	Design, management

The underlying causes, failure to identify the risk, were design failures (overcapacity, overcrowding, too many people, not enough space), then information failures (where changes in the environment, such as handouts, change in weather, false alarms), then management failures. We also identified the phases of crowd failure, during ingress, during circulation (mid event) and during egress.

We called this the DIM-ICE risk analysis model.[15]

DIM-ICE RISK MODEL

We developed the DIM-ICE risk model from the analysis and classification of over 100 years of accidents and incidents in crowded spaces. It was based on identifying the principle, underlying (distal) causes of crowd-related disasters. Three common categories of failure emerged – design failures, information failures and management failures. We can, of course, break down failure modes into several different sub-categories, but the top-level (design failure, information failure and management failure) serves to focus the user on phases of crowd behaviour and influences on crowd behaviour. The aim was to force the user to think through both the phases and influences on crowd behaviour, to then formulate a crowd management plan that systematically progresses through the phases of crowd movement (ingress, circulation, and egress) while addressing the influences on that behaviour.

We excluded the incidents that were unexpected and catastrophic failures, such as terror attacks, fire and structural failures. Structural, electrical and construction elements have their own type of risk assessment typically based on testing electromechanical systems to destruction. Building code compliance and structural/electrical standards govern those types of risks. Our focus was on the preventable, crowd dynamics, failures.

We did not include terrorist attacks on the crowd. However, later in this chapter, we highlight a paper that does include attacks by Mark Breen, one of the MSc students.

PROXIMATE AND DISTAL

Breaking the accidents/incidents into the primary distal (underlying) and proximate (immediately before the incident) to identify the principles that we should be looking for to prevent future incidents, the classification broadly fell into three catagories: design, information and management failures.

[15] DIM-ICE, because of the similarity to the word demise, death.

Design failures: The types of incidents that were due to the volume exceeding the capacity of the system. There were recurring elements of failure due to overcrowding, overcapacity and throughput, where the volume of crowd flow exceeded the capacity of the route width, where more people were trying to access the space that was safe to do so.

Information failures: The types of incidents that occurred when the environment changed. Such as an unexpected change in the weather, a last-minute goal and political rallies where an unexpected celebrity appeared. These fell into the classification of information influencing the crowd behaviour.

Management failures: Where the crowds were mismanaged or reacted to the management process (such as corralling or heavy-handed stewarding causing the crowds to react to the people mismanaging them).

PHASES AND INFLUENCES ON CROWD BEHAVIOUR

We identified three primary influences on crowd behaviour design, information and management. There are also three primary phases of crowd movement, ingress, circulation and egress. Crowd behaviour is different in each of these three phases.

Ingress: When people come to an event, a football match or concert, what are they looking for? The way to the event, the way to their seat, and time, the start of the event, motivates people to get in, and get the best position to vide the event, and/or get to their designated seats, and get comfortable as soon as possible. Ingress behaviour is influenced by the pre-event information, the signage, and the site design (routes to and from the event/venue).

Circulation: Crowd behaviour mid event may be quite different from the crowd behaviour during ingress/egress. The crowd is already in place, within the event/venue. Circulation may involve crowd dynamics, at say, halftime, or an interval at a concert or sporting event, when people are looking to how to get from their seat to the toilets, to the food and beverage counters and to and from the merchandising areas. Again, the behaviour is influenced by the signage, information systems and local geometry of the site/venue.

Egress: The crowd just wants to get away out of the event/area and away from the site as quickly as possible.

COMPLICATED

Now the dimensions of the problem are beginning to look complicated. We have just defined three distinct types of behaviour during three separate phases of crowd movement, *and* we need to understand how these phases and influences affect crowd behaviour under both normal and emergency situations. At the start of this book, we outlined a suite of *simple* risk assessment tools, and we will get there. However, we do need to lay the foundations of the problem, so you can understand the contributing factors of crowd disaster, and a method of seeing through the lens of past disasters. The aim is to focus your attention on those key elements of past causality and to visualise those elements for your own sites/venue/events before the crowds arrive on your site. Let us first expand the dimensions of the DIM elements of the crowd dynamic.

DESIGN-INFORMATION-MANAGEMENT

If we consider the site design, the areas people move and occupy, we need to understand how the crowd occupies different spaces. We typically define crowded spaces as either static or dynamic—static, where crowds are stationary, and dynamic, where the crowds are moving. There is a third category which is a contingency space, those spaces that we need to set aside in the event of an emergency either to decanter the crowd, or to gain emergency access. Therefore, the routes that we can move a crowd are important to understand. Ingress and egress flow rates under normal and emergency conditions will be an essential dimension of our crowd risk analysis.

Information refers to signage, maps, social media, news reports, site communication systems, again under normal and emergency situations. For example, how do you communicate with the crowd, how do you influence the crowd's behaviour and how do you shape the crowds' expectations? Again, we can draw on the volumes of work published by illusionists and magicians. For example, in *Magic in Theory* by Lamont and Wiseman, there is a summary of directing perceptions;

> Physical misdirection directs where and when the spectator is looking by using and creating areas and moments of primary and secondary interest. The spectator's attention can thereby be diverted from the area of method and reduced at the moment of method. His attention can be diverted towards the area of effect and increased at the moment of effect. By controlling the spectres attention, the magician can direct what the spectator perceives.

We have surveyed and advised on signage at sites/venues/events over the last few decades, and, in general, site signage and information systems are inadequate. We have seen examples where signage is barely visible (useless) and signage conflicts with local information (confusing). Is it the magician's "eye" that allows us to see so many site issues? It is not just the magician who understands how the mind perceives, for example, the difference between saccade[16] and fixation eye movement, how to induce inattentional

[16] http://www.scholarpedia.org/article/Human_saccadic_eye_movements

blindness, how to shape the audiences' expectations to enable the suspension of belief. The same principles are the science of traffic signage.[17] The same principles apply to shopper[18] information systems.

Finally, management is not just the number of security guards/stewards or ushers you have in place. It is your process and procedures, your communication system, your chain of command under normal and emergency situations. How your team reports or red flags' issues or problems on site, during ingress, during circulation, and during egress? How do you identify if things are still going to plan?

INGRESS-CIRCULATION-EGRESS

During the primary phases of crowd behaviour, ingress (getting in) may require a queuing system, of barrier design. You need signage or an information system. You may need to manage the flow of information if there are delays or problems on the site. You may need to inform people heading towards the site to direct them away if there is an emergency. You will need to understand your flow rate over time, and you need to detect problems as they are starting to arise. Red flagging a problem during ingress should be part of your management reporting process.

When people move around your site/venue during circulation, how do you red flag any problems? You need to know what the crowds are expecting and how to manage those expectations, for example, the appropriate allocation of male/female toilets, their location, the concessions, food and beverage and merchandising your onsite signage. All of this (and more) will shape the crowd behaviour and crowd movements during circulation around the site/venue. How do you red flag any issues? What are your reporting processes and procedures to identify a manager or supervisor that you see something you are not expecting?

During egress (getting out), you will need to assess your crowd flow rates for normal and emergency situations. How do we know the limits, the maximum crowd flow off a site/venue? How do we know if that flow or capacity limit is beyond a safe limit? How do you red flag any problems during egress?

ANTICIPATE, PLAN, PREPARE AND PREVENT

This is one of the first learning points on the course, your crowd management plan. Typically, these will have a section "emergency evacuation", which may (typically) read as follows:

---

[17] http://www.signresearch.org/wp-content/uploads/Arts-and-Science-of-Sign-Design.pdf
[18] https://explorerresearch.com/effective-signage-design/

> In the event of an emergency, the entertainment will cease, the house light will come on and everyone should leave the building/venue/site by the nearest available emergency exit, in an orderly manner.

Which is a mid-event evacuation procedure? What if the emergency occurs during ingress, or during egress, when the crowds are on the move? When we examine most accidents and incidents, they occur during ingress, or egress, when the crowds are entering or leaving a site/venue/event? An emergency *during* ingress or egress will have a different crowd dynamic than a mid-event emergency. The latter is simply when the alarm sound stops everything, and everybody leaves.

Suppose the crowd is heading towards a venue, and you discover a situation where you need to clear the people away from the venue/site/event. In that case, it may be partially full, so we have an additional dimension to the problem. You might need to get out and you still have people heading towards your area. That crowd may need diverting away from the site/venue/event.

You may have to deal with patrons, having paid for a ticket trying to gain access, and you have people inside. It would be best if you got out. This is a vastly different and two-fold communication strategy – informing the approaching crowd to stay away and informing the crowd inside your site/venue/event to get out.

What happens if the crowd starts to leave, and you discover a problem during egress? It could be a suspect package on an egress route, a medical emergency, a scene of a crime. Again, you need a strategy of communicating with the crowd, to divert the egressing crowd, to hold them in place.

If your crowd management or event plan does NOT have a section outlining your emergency procedures for an incident during ingress emergency, AND a section outlining emergency during egress, if it ONLY has a section outlining your mid-event emergency, then it is simply not fit for its intended purpose. Without a plan in place to cover all three phases of crowd behaviour, you have failed to anticipate, failed to plan and hence failed to prevent a situation from escalating into an incident.

OVERLY COMPLICATED

This is the first lecture in the course, one of the first chapters in this book, and the problems of crowd risk analysis and crowd safety are now looking *overly complicated*. We promised simple tools, easy to use, easy to understand and easy to apply. We need to define a common framework for all events, for all crowd risks, and to apply a set of tools to any site.

How do we do that, surely all events have unique characteristics, how can we use one set of tools for all sites/events/venues. Is this even possible?

Table 2.2 The DIM-ICE risk model – blank template

Normal	*Ingress*	*Circulation*	*Egress*
Design			
Information			
Management			

Emergency	*Ingress*	*Circulation*	*Egress*
Design			
Information			
Management			

TEMPLATES

Over the last three decades, we have taught and applied these principles to the planning, licensing/permitting and management aspects for places of public assembly. This process requires a brief explanation. To start with, the DIM-ICE risk model is a blank template (Table 2.2). By driving the user to think through the details that need to go in each box, we force the user to *think* about the influences on crowd behaviour.

The method (filling in the blanks) is the essential element in this process. If you have a blank box, you have a gap in your knowledge. So, the filter of considering the design for ingress (normal and emergency situations) forces the user to consider the emergency during ingress – similarly, emergency during circulation and egress. How do we inform the crowd, how do we manage that phase of the process? The objective was to force the user into thinking about how to influence the crowd: design, information and management are the primary influences, during both normal and emergency situations for ingress, circulation and egress. It is also a template for writing your crowd management plan.

STRUCTURING A CROWD MANAGEMENT PLAN

Chapter 1: Ingress
- Design (for normal and emergency situations)
- Information (for normal and emergency situations)
- Management (for normal and emergency situations)

Chapter 2: Circulation
- Design (for normal and emergency situations)
- Information (for normal and emergency situations)
- Management (for normal and emergency situations)

Chapter 3: Egress
Design (for normal and emergency situations)
Information (for normal and emergency situations)
Management (for normal and emergency situations)

The structure of crowd management plans vary from the event organiser to event organiser. There are no standards and we have seen everything from the one page to supply stewards, barriers and a skip for £/$ to volumes of complete detail. That is part of the problem with crowd safety management, the lack of standards for documentation. However, for an accident/incident investigation, there are key documents that will be examined during litigation. One document would be the sites/events/venues risk assessment, another the crowd management plan and another the licence/permit application documents.

In our experience, it is obvious when there is a significant amount of "cut and paste" from previous documents. We would normally compare this year's crowd management plan and risk assessment, with previous versions produced for those sites/events/venues. We would look for the similarities and differences from this year's plan to previous plans, from this site's plan to other site plans produced by your organisation. It may come as no surprise that legal cases can win or lose on the documentation quality, specifically the risk assessment. The critical prosecution questions of whether the event organiser recognises the risks at this event? And if those risks are documented in the event/crowd management plan.

SPLIT THE DOCUMENT

We advise our delegates to split their event documents into sections. There will be sections that are the same for each event, with minor changes from event to event, such as company details, contact names, management structure, terms and conditions, process, procedures, etc. There will also be unique elements to each event, such as site plans, maps, diagrams, etc. If you mix this information into one "master document", you will need to update each fragment in the whole. It is simpler to update *just* the site-specific data. So, separate these sections, stuff that is the same from event to event, and stuff that is unique to this event, into different documents, so that you only need to update the individual elements. It is much less work, and you can focus on the unique information on that site.

USE THE MODELS

Modelling the site using DIM-ICE and other tools (later in this book) will help your document, make it more accessible to the licensing/permitting and /or multiagency teams reading the document. Our tools are visual in nature and designed to maximise the information while minimising the amount of work to process the information. Again, the objective was to encourage a structure to both the risk analysis and understanding of causes of accidents and incidents, to create a summary (in the table) that leads to a structure in an event plan. The next phase of the process is to highlight the elements of the DIM-ICE model, using Green (things we think we do well), Orange/Amber (things we need to monitor) and Red (things we need to improve).

Two DIM-ICE examples (Table 2.3), from client projects and illustrate how the DIM-ICE model captures information and presents it in a way that provides a team input. For example, "We think this box should be amber, we need to communicate this to the operators of the site, licencing, and planning, and communicated our concern in a visible table, to the budget holder. This is where we need to invest resources and funding".

DO IT IN PARTS

In general, we do not apply the DIM-ICE model to an entire site. We break the site into sections or parts, for example, an entrance or an exit. Then we can combine DIM-ICE models, and their information, on a site map and we see additional dimensions of the crowd dynamics, and crowd risks, across the site.

APPLICATION – CANADA DAY

In 2016, we ran a series of workshops in Ottawa relating to the use of the various risk models. Kevin Kennedy outlined the core issues of the site during the workshop.

> ### Canada Day – General History – Kevin Kennedy[19]
>
> Historically, the focus of Canada Day has been on Parliament Hill. This location is the venue for the Changing of the Guards, speeches by the Governor General and the Prime Minister in addition to the live performances for the noon and evening shows. After the evening show, fireworks are launched to illuminate the skies behind and above the Parliament Buildings. There are other areas with attractions (Majors Hill Park, the Cenotaph, Sparks Street, and the Byward Market) but the Parliament Buildings has always been a main destination for the crowds. Canada Day is a free event for all visitors and for decades people moved freely from site to site.

[19] Kevin Kennedy is a Staff Sergeant with the Ottawa Police Service (OPS) and has over 36 years of policing experience including 15 years with the Ontario Provincial Police (OPP). His introduction and his essay features later in this book.

Table 2.3 Two examples of a DIM-ICE model

Normal	*Ingress*	*Circulation*	*Egress*
Design	North approach 8 m, Public Transport not available – walk up, 7 m entry	Wide event area, four video screens to draw people away from official area.	7 m roadway blocked to allow people to egress safely.
Information	RSL website commence 2 months, prior, social media, news media education	Information signs, police, SES, security ushers, park and ride available. Sand may challenge and aged and disable.	Signage and usher support on egress. RSL website education provided.
Management	Traffic management, Plan implemented, Road closures Signage and volunteers to guide Speed limit reductions	Challenge to move people around venue due to environment. Veteran parade area to be clear. Difficult to move from N-S in the event.	Sudden movement post event. SES, police direction, security, traffic controllers and transport hub personnel.

Emergency	*Ingress*	*Circulation*	*Egress*
Design	Emergency services on site. Access for additional emergency vehicles. Ambulance and first aid teams positioned around area.	Sand and fencing restrict easy movement and prevents lateral movement. Official area restricts North – South flow.	Additional emergency egress routes available along beach and via closed central roadway.
Information	Clear signage and guidance from officials. RSL website to be used to educate patrons.	Signage difficult due to venue design. Moving on sand and roadway difficult. Size of area difficult to deliver messages.	Signage difficult due to venue design. Moving on sand and roadway difficult. Size of area difficult to deliver messages.
Management	Signage, barriers, Security, police No vehicle. Open roadway	Police commander to provide direction to all emergency service personnel through EOC.	Briefing and mapping of evacuation sites prior to event for all staff. Directed evacuation. Beach area to north and south to be used.

Normal	*Ingress*	*Circulation*	*Egress*
Design	Visible and clear entry points. Space to queue at entry gates. Capacity for 3,000. Pram parking areas Narrow stairs.	Narrow stairways. Pram and wheelchairs blocking concourse flow. Competitors' area fenced off and patron access restricted (and axes secured)	Narrow stairways. Six exit gates. Egress into crowded areas.

(*Continued*)

Table 2.3 (Continued) Two examples of a DIM-ICE model

Normal	*Ingress*	*Circulation*	*Egress*
Information	Show maps. Schedule in daily papers. Social media alerts. Directional signage. PA announcements	PA announcements. Ushers visible. Large internal screen. Internal signage.	PA announcements. Ushers visible. Internal large screen. Internal signage.
Management	Morning briefings with emergency services and internal staff. Maintain queuing space. Security team roaming. Competition program and wet weather schedule. First aid response onsite.	Communication with team members. Queue management. First aid Response onsite. Control room (monitoring external crowd flow via CCTV). Security teams roaming.	Collection of prams. Egress into crowded areas. Ushers assisting at stairs. First aid response onsite. Control room (monitoring external crowd flow via CCTV) Security teams roaming.

Emergency	*Ingress*	*Circulation*	*Egress*
Design	Crowding on Grand Parade and Monaro Street. Capacity for 3,000. Visible and clear entry gates.	Narrow stairways. Prams and wheelchairs blocking concourse flow. Competitors' area fenced off and patron access restricted (and exes secured).	Narrow stairways. Six exit gates. Egress into external crowded area.
Information	PA announcements. Emergency management plan. Communication with control room. Internal large screen. Internal signage.	PA announcement. Emergency management plan. Communication with control room. Large internal screen. Internal signs in position.	PA announcements. Emergency management plan. Communicati9on with control room. Large internal screen. Internal signs in position.
Management	Ushers visible. Communication with control room. Emergency services on site. First aid response onsite. Additional security response teams available.	Ushers visible. Communication with team members and control room. Enforcement of competitors' safety procedures. First aid response onsite. Emergency services on site. Emergency services on site. Additional security response teams available.	Egress into a crowded area. Emergency services on site. Additional security response teams available. Communications with control room.

Impact of International and Terrorist Activities

Since the start of the millennium, terrorist incidents have become more prevalent around the world. In response, security around Parliament Hill has tried to keep pace. Metal detectors were installed just inside the building entrances which slowed the ingress of visitors. In 2012, Public Works made the grounds more secure by installing retracting bollards at the road access points and narrowing the pedestrian gates along the Heritage Wall. The size of the Queens Gate was not altered, but it remained closed to the public as it only used by the Royals during visits. The narrowing of the pedestrian gates did influence crowd flow, but people were still able to move freely on and off the Hill during Canada Day.

On October 22nd, 2014, there was a terrorist incident at the Cenotaph which transitioned to the Parliament Buildings. In the aftermath of that incident The Parliamentary Protective Service (PPS) was established. Rules regarding access to buildings were changed and access to the outdoor grounds became more restrictive during large scale celebration like Canada Day. Visitors now must pass through security screening checkpoints at designated areas before they can access the Hill grounds. These check points were initially established at key gates, reducing the overall number of access points through which crowds could pass. This not only slowed the pedestrian speed, but also reduced the cumulative width of passageways which further slowed crowd movement.

Canada Day Crowd Concerns Prior to the Sesquicentennial

It was required that all security screening had to be conducted on Parliamentary Precinct property (all property north of the Heritage Wall which includes the Great Lawn in addition to the East, West, and Centre Blocks). This meant that the security screening tents had to be positioned just inside the Heritage Gate Wall and could not be located on the Wellington Street sidewalk or roadway. During the 2015 and 2016 Canada Day celebrations, overcrowding was experienced on Wellington Street caused by the positioning of the security screening tents, the limited gate widths and the lack of crowd management infrastructure on the city side of the Heritage Wall. This lack of control contributed to crowd congestion, crossflows and counter-flows which increased the overall risk to the public. Furthermore, the morning barricading along the Royal Route (positioned to protect the Prime Minister's and Governor General's arrival at the Hill) further restricted crowd movement. It was identified that potential crowd risks needed to be better mitigated prior to Canada's 150th Canada Day celebration in 2017. In 2016, Dr Still was consulted and he presented a short course to key city planners and then further assisted with some recommendations to better address crowd issues.

ADDRESS THE KEY SITE ISSUES

Kevin provided a comprehensive overview of the site problems during the crowd science workshops in Ottawa, with the Royal Canadian Mounted Police, Heritage Canada, Ottawa Police, OC Transpo, and several other agencies involved in the planning and management of the event.

In the UK, there is a process called a TENS[20] (Temporary Event Notification) used to grant various licences for space, in which you want to host an event. The problem we noted in Ottawa, with Parliament Hill, was the design limitation. By breaking the entry/exit gates into the DIM-ICE models (shown in Figure 2.1, North is up) below, we could identify the pinch points and assess the throughput against a range of potential crowd sizes.

This led us to review the site design. In all entry/exit points, the architecture/gates were the site restrictions. This reduced the movement of crowds on and off the site. There is simply not enough space to accommodate both normal and emergency situations. The use of the DIM-ICE model (Figure 2.1) positioned for each of the gates on the site, highlighted the critical problems of overall site ingress/egress. Figure 2.2 shows the gate, and counter terrorism barriers, which impeded the crowd movement.

Figure 2.1 DIM-ICE risk model applied to a site.

[20] https://www.gov.uk/temporary-events-notice

Figure 2.2 The central gate to Parliament Hill.

Kevin Kennedy – Identified Concerns for 2017

150th Anniversary – there was no way to predict the number of visitors that would attend but it was expected that this milestone Canada Day celebration would attract the largest crowd experienced to date.

Royal Visit

It was highly anticipated that a representative of the Royal Family would make an appearance on Canada Day. Given the popularity of the Royal Family with Canadians, this would further attract more visitors who would not otherwise attend.

Potential for terrorist activities

There was a heightened awareness for the potential for harm due because of the significance of the event, the general open nature of all events in Canada, the presence of VVIPS, the international attention that was already focused on our celebration and the fact that in the past Canada has been highlighted as an enemy by terrorist organizations.

Previous Crowd issues

As mentioned, the positioning of security screening tents having to be on Parliamentary Precinct property have caused issues. The design has created queue lines on Wellington Street that impacted other crowd flows. With the larger crowds anticipated, the congestion issues would be more so. In addition, the barricading and extra security along the

Royal Route for Prince Charles would only exacerbate crowd congestion issues further.

Emergency Egress from the Hill

With limited access routes on and off Parliamentary grounds, emergency egress is always a concern particularly when some of those routes are already congested by people lined up to enter. A solution needed to be identified that would allow for a mass egress in the most expedient and safest method possible. Having the capacity of the Hill (~35,000 people) leave quickly through the multiple 1.5 m gates along the Heritage Wall (some routes congested with people lined up to enter the Hill) is a potential risk that needed to be addressed.

The ingress/egress problems are more pronounced in the East, where most people arrive/depart due to the location of the mass transit system. Site ingress/egress is the problem, and this is due to the gate width. We cannot alter the historical architecture, but need to increase the overall crowd flow on/off the site, so how do we change the design?

THE GRASSY SLOPE

There was a grass slope on the Parliament side of the fence, and that tapered from the height shown in Figure 2.3, to level with the road. Wellington Road sloped down from West to East.

Figure 2.3 The ornamental fence and the grass slope.

Figure 2.4 The stairs over the fencing on Parliament Hill.

OH NO!

One of our children's favourite bedtime stories was *We're Going on a Bear Hunt*[21] written by Michael Rosen and illustrated by Helen Oxenbury, in which the family encounter a range of obstacles with the mantra,

> We can't go over it. We can't go under it. Oh no! We've got to go through it!

This struck a chord for the Parliament Hill ingress/egress problem. We could not go under it or through it, but could we go over it? The gates (Figure 2.2) were too narrow to accommodate the anticipated crow flow, but we could build a set of stairs to go over the fence. This would significantly increase the crowd flow, one and off the site, under both normal and emergency conditions (Figure 2.4).

However, this turned Wellington Street into part of the event space. There was an issue about the powers to stop and search patrons on the streets. It was Ottawa Police territory. The authority to stop and search as patrons entered Parliament Hill fell under the authority of the PPS[22] (Parliamentary Protective Service). By extending the event footprint, we shifted the authority to stop and search. Kevin continues his narrative of the solution.

[21] https://en.wikipedia.org/wiki/We%27re_Going_on_a_Bear_Hunt

[22] The Parliamentary Protective Service (PPS; French: Service de protection parlementaire) is an entity of the Parliament of Canada created by law that mandates the physical security for parliamentarians, employees, and visitors within the Parliamentary Precinct in Ottawa, Ontario.

Kevin Kennedy – Implemented Solutions

The restrictive principle for Canada Day is that PPS was required to screen the public before they accessed Parliament Hill. However, PPS could only conduct that screening on Parliamentary Precinct property. Staying within those restrictions in 2015 and 2016 required that the screening tents be positioned just inside one or two gates. This inevitably created crowd issues on Wellington Street. This also had the added effect of reducing the available emergency egress routes increasing risk.

The solution

Make the Parliamentary Precinct bigger. In doing so you could move the screening tents further out from the Heritage Wall and PPS would still be able to conduct their screening on Parliamentary Precinct property. Furthermore, once screened, people could move freely on and off Parliament Hill without restriction. Additionally, the restrictive nature of the gates could be overcome by building large structures over the walls with staircases to the street level. This would significantly increase crowd flow which is most important during any emergency egress.

Issue

This would require an Act of Parliament. Several discussions took place with multiple stakeholders and the Act was passed. The infrastructure was established over a 3-day period prior to June 30th and was used for the three-day celebration.

We prompted an act of Parliament that would make Wellington Street, a temporary extension to the area of Parliament Hill. We could now position the security checkpoints on Wellington Street, where there was sufficient space to manage the crowds, under the authority of the PPS.

The DIM-ICE models identified the nature of the site problem; the capacity of the gates was less than the anticipated crowd flow. The solution was to increase the capacity to meet the demand, we could not change the existing gates, but we could increase the capacity by building a temporary structure over the fence line. The revised DIM-ICE models then showed green for the design, both normal and emergency, we have increased the capacity.

CRITICAL ANALYSIS

When teaching, specifically at the MSc level, we always encourage our students to challenge the existing knowledge and publications. Developing critical analysis skills is both essential for improving knowledge and understanding, and a university requirement. We also encourage students to assess our own models, to improve them, or confirm them as valid, and to develop their own models to add to the body of work for the crowd sciences.

The next two sections are papers by MSc students on the Crowd Safety and Risk Analysis course at Manchester Metropolitan University. The first is a critical analysis of the DIM-ICE model and the second proposes a different model, based on other disaster analysis.

As we have stated in this chapter, we had omitted attacks on the crowd as those are external to the control of the safety team. Assessing external risks, specifically of a terrorist attack, is a different filed (counter terrorism) and there are experts on those areas of crowd safety. The second paper looks at the "attack" risks and outlines a useful classification for the event organisers risk assessment.

Steve Riches – DIM-ICE: A critical analysis

About Steve

Steve has been working in the field of event safety for over 10 years having started as a safety advisor for Leicester City Council, cutting his teeth on events such as the Leicester Diwali, Caribbean Carnival, Riverside and Summer Sunday Festivals. The passion for making events a safer place to enjoy the outdoors led him to undertake the Event Safety Diploma at Derby University which further drew him in to the world of emergency planning, structural safety and crowd safety. It was clear to Steve having seen the massed crowds at large events that there had to be a better way to analyse and assess the hazards and risks of crowd movements, whether in the open or in complex-built environments.

Steve first met Prof. Still at the Emergency Planning College where he completed two short courses in Crowd Dynamics which was a turning point for his work in crowd safety and risk analysis and this gave him some confidence to take on larger projects to gain more experience. Steve went on to be one of the Health and Safety Advisors to LOCOG in London 2012 and led the health and safety element of the Common Domain which regularly had crowds of 58,000–200,000 people a day wandering through those areas. From here Steve has provided key safety advice and support to other large events including the Glasgow Commonwealth Games Festival and the Inaugural Games in Baku, Azerbaijan, before leaving for sunnier climbs for three years in Abu Dhabi at Yas Marina Circuit.

Latterly Steve has been undertaking studies at Manchester Metropolitan University where he has been able to draw on his empirical experience to critically analyse the key fundamentals of planning for crowd safety and analysing the theories and tools behind what he witnessed in the previous years. Steve now works for A.C.T. National Ltd in the UK and provides specialist support and advice to organisers who need to plan events with large crowds and helps analyse the dynamics of crowd movements, one of the key tools used for this is the DIM ICE and RAMP Analysis he first learnt at the Emergency Planning College. With his expertise in fire safety management, emergency planning and control room management, Steve can draw on his academic learning to critically evaluate escalating issues in real time and provide the appropriate response before the matter becomes a major problem.

Introduction

This paper looks at the Design, Information, Management – Ingress, Circulation, Egress tool (DIM-ICE) and assesses the techniques used and any limitations to the techniques and have been compared to other safety tools such as FMEA and Quality tools such as Ishikawa Fishbone model. It is established that DIM ICE is a tool that provides subjective outcomes from objective data entry, but this is not controlled to final reaction. There are some weaknesses to DIM ICE which are explored; however, conclusions discuss combining DIM ICE, Quality and risk assessment as one meta model, thereby taking more control of the planning process.

The focus of this analysis will be upon the use of the crowd assessment tool known as DIM-ICE for the purposes of identifying crowd dynamics, behaviours and design issues to reduce the risk of harm to heterogeneous and homogeneous crowds that are either static or dynamic. The tool is an aid to the risk assessment that must be undertaken to identify hazards and reduce the risk of harm.

There are several tools used in safety that undertake the same sorts of processes in design and planning, however, as will be seen by analysis, there is no single tool that can be used from conception to delivery that not only provides a systematic process of the planning stages of an event but also capable of being used by lay people to identify those dynamics they should reasonably be aware of. For a tool to be pragmatic to the event organiser it must be able to quantify the level of risk identified in order that an action can be assigned, too often there are issues that arise that people do not react to early enough because they either do not understand the risk or it is not brought to their attention in a format, they are able to understand and therefore prioritise.

Process safety and quality control especially in the manufacturing and chemical world are as close to traditional safety as crowd dynamics get, very often one can look at the movement of people in the same way scientists look at the flow of water or the breakdown of a chemical reaction.

The DIM-ICE risk model is broken down in to three primary phases of crowd behaviour, three primary influences on that crowd behaviour and across two primary nodes of crowd behaviour [Still, G. K Crowd Safety & Risk Analysis [Online] [Accessed 30/09/17] http://www.gkstill.com/Support/WhyModel/dimice.html

In other tools such as the Ishikawa fish bone model, a similar structure is utilised in that there are six measurable elements that if not controlled would lead to a defect, however, as with other safety tools such as Failure Mode Event Analysis, Fault Tree Analysis, Hazops and Event Tree Analysis, we will see that these do not easily fit the crowd dynamic model even though they are able to provide quantitative data to justify outcomes due to the number of variables in behaviour that are not subject to precise control.

Analysis

At the conception of any event, it is generally intended that people visit the event for business of leisure. This can be in low or high numbers and depending on the design and attraction of the main event, will either be an event with areas of low or high densities. To reduce the risk of harm to those attending an event it is critical that the organiser can fully appreciate the impact the crowd could have on the event, even when all is going well. Understanding what influences crowd behaviour is vital to ensure that the plans are set in motion to manage and if necessary, take control of certain situations.

DIM (Design, Information, Management) – To start with, the design of an event will have a huge impact on how the crowd moves around site, if the crowd is properly informed and is able to use all five senses to make informed decisions on their own, it will reduce the burden on management who can then get on with other tasks and lighten the mood of the crowd. The mood in this sense will be understood to be a function of the event and everything to do with the mood is intertwined with the event such as the environment the crowd is in, the demographics of the crowd and the entertainment that is the event.

Aspects of the design that affect the crowd include but are not limited to such matters as the weather, demographics of those around or in the crowd, traffic, music and other entertainment, unexpected incidents. In addition, where flow is concerned the design of the event will affect how people move.

ICE (Ingress, Circulation and Egress) are primary phases of an event that must be recorded in conjunction with the influences of a crowd. A crowd will always arrive and leave an event, however whilst a lot of work is put in place for access and egress, this tool allows the user to remember to plan for circulation. Advanced users of the tool have also added additional phases in this section to include arrival and dispersal.

Using the tool has its challenges however, although it is a simple matrix, it is data hungry and therein lies a key weakness. When data are late or there are numerous others inputting data, it is possible that the information needed to make a reasonably informed decision about a control, or the level of risk could also be delayed. The way this is commonly used is to input the data in the tool as the task should be optimally operated. If instead the tool mentioned a snapshot in time, either from plans or observations, it could be used as a tool to the risk assessment that should accompany the matrix.

DIM-ICE is a qualitative assessment and, in most cases, based on people's subjective assessment of the findings entered. When combined with a quantitative or objective assessment that can be rated, it becomes a powerful tool that must be reacted to. Although the tool is not recognised in traditional safety as a risk assessment, if it were not to be used at an event where there had been a crowd incident, there is a high probability that the courts would take a dim view, as this is also seen as industry best practice.

A difference between DIM-ICE and general risk assessment is that there is no real necessity to provide numbers to the data entered. The use of colours to identify where the author considers the risk to lie is as close as it gets to instructing a reader to act. The problem however is that there is not the general competency for people to be able to undertake a general risk assessment of crowd management, and therefore they generally do not get undertaken. The DIM-ICE model as a tool that aids risk assessment has an additional feature that does not work anywhere else in either traditional or more modern safety systems, the fact the model has identified two modes of operation, normal and emergency means that there is a good chance that any potential issues on a route can be identified early and considered for action prior to becoming a problem. It is a well-known fact that an event in an emergency has its own management approach which can be quite different to normal run operations and this model allows for that level of planning intervention to take place.

Other tools such as an Event Fault Tree requires an initiating event for it to work, if that initiating event is a failure of a process resulting in injury, then it would be the wrong tool to use for crowd dynamics. A Hazop used in the chemical industry is the closest tool to DIM ICE/ Risk Assessment as it allows for a systematic examination of a complex process or design to evaluate problems, and it is extensively used to pick up design issues. In addition, the language of the outcomes, consequences and likelihoods of the nodes are similar to general risk assessment, so are very easy to learn [Hazard and Operability Study[23] Accessed 30/09/2017].

If it is not risk that should be measured in terms of crowd management then instead there may be an argument to consider crowd management as a quality control of an event. There are several major tools

23 https://en.wikipedia.org/wiki/Hazard_and_operability_study

in manufacturing including 5S and LEAN that use such tools as the Ishikawa model. A tool from the 1968s to reduce defects and prevent imperfection, where there is such imperfection, it would be seen as a variation from normal taking into consideration the whole process by major element category of which there are six. To complete the assessment, it is then rateable against a set criterion and can therefore also be used as part of a root cause analysis should something go wrong. [Ishikawa K 1990, Introduction to Quality Control (translated by Loftus, JH) 448p]. This is the closest model to DIM-ICE that also has the capability to consider both the objective and subjective nodes, rate them and give actions a priority to improve elements of crowd management at an event.

Conclusions

There is no doubt that a tool is required for Crowd Dynamics, and since 1999, DIM ICE has provided a tool that aids the process of risk assessment. The fact that others do not conduct risk assessments is a criminal liability not understood by many event organisers. It is evident that there are other tools in safety that need to be properly reviewed to establish if they can offer any improvements to the range of tools that are available. It is contentious perhaps that there is only one tool available across the world that is known to provide crowd dynamic information for risk assessments and perhaps, even more arguably it should be less subjective and more objective by incorporation of a quality model alongside. As with all safety tools, there is no point giving information to people who know not how to react to it. If it were possible to combine DIM ICE model with Ishikawa principles and provide it in a format that a court would accept as a risk assessment, it would allow event organisers to take control of a part of an event that up to this time is rumbling under the hood of the car. Crowd safety is a critical element of an event, and at present, it is a DIM ICE assessment away from bursting its cam belt!

References

Riches, S., 2011, Principles of Risk Assessment: White Paper Siemens PLM Software. How to conduct a Fault Mode Event Analysis: 2016: White Paper [Online] [Accessed 28/09/17]: https://polarion.plm.automation.siemens.com/hubfs/Docs/Guides_and_Manuals/Siemens-PLM-Polarion-How-to-conduct-a-failure-modes-and-effects-analysis-FMEA-wp-60071-A3.pdf

Redmill, Felix, Risk Analysis – A subjective Process : Redmill consultancy: [Online] Accessed 30/09/2017] : http://homepages.cs.ncl.ac.uk/felix.redmill/publications/5D.Risk%20Analysis%20a%20subje.pdf

Sime, J.D., 1995, Safety Science 21: Elsevier.

Still G. K., 2012, Understanding Crowd Dynamics: Lecture Notes.

Still G. K., 2012, Mood of the Crowd: Lecture Notes.

Mark Breen – The LaD Classification Model for Crowd Disasters

About Mark

Mark Breen is a Risk and Safety Management professional, specialising primarily in Crowd and Event Safety. He is based in Dublin, Ireland, where he is a Director of Safe Events. Mark and his Team work domestically and internationally with clients in the Middle East, the UK and Ireland mainly.

Mark operates both strategically and operationally, often filling HSE(Q) Director roles on large international festivals as well as on-the-ground Crowd and Event Safety roles for a wide variety of clients. He is a regular speaker, commentator and contributor on topics around Crowd and Event Safety and is extremely active in the global events community. Mark is heavily involved in the learning and education space and is passionate about doing whatever he can to help people run safer events.

His hope is that he can have a positive impact on events that he will never even know about, through developing and delivering courses and educational resources, etc., in parallel with his hands-on work in the field.

Abstract

Crowd disasters are still commonplace, and the vast majority can be avoided. People are killed and injured needlessly all around the world at events where crowds gather. The environments in which crowds gather around the world differ significantly; legislatively, developmentally, infrastructurally, religiously, experientially and in many more ways. This understanding has led to the development of the proposed LaD Classification Model for Crowd Disasters.

The model classifies disasters based on whether they happened because of a lack of elements key to the safe planning and management of crowds or a disregard of elements that did, in fact, exist. There is also an attack classification, which is assigned to disasters that occurred, in the first instance, because of a planned and executed attack on a crowd.

It is concluded that the use of this Model can inform the efforts of those seeking to improve the safety of crowds the world over – be they Governments, police forces, private operators, academics, or any other interested party – by aiding in identifying the areas that should be focused on for development / improvement. It is further concluded that the inclusion of the Attack classification will act as live barometer of the increase in attacks on crowds as an issue affecting those in charge of crowds worldwide.

Background

As noted by Fruin in Causes & Prevention of Disasters (1993),

> Most major crowd disasters can be prevented by simple crowd management strategies.

Fruin is correct in that most of the Crowd Disasters that have occurred have been preventable. A simple set of crowd safety tools and techniques that could have prevented many of these Disasters, already exists. However, more incidents could have been prevented if existing regulations and legislation had been implemented / or enforced.

In researching the Disaster Database[24] there are two broad categories by which the Disasters can be classified, and this has driven the development of the LaD Classification Model for Crowd Disasters.

The LaD Classification Model for Crowd Disasters has developed from previously unpublished work by the author. This classification Model was previously termed the DoD Classification for Crowd Disasters – Dearth or Disregard – with the focus being solely on whether those in charge of the event experiencing the disaster were (a) suffering from a dearth of training/understanding/regulation/compliance-enforcement, etc., or guilty of a (b) disregard of regulation, enforcement or compliance orders, laws, expert reports, available expertise, or knowledge, etc.

With the proliferation of attacks on crowds in recent years, the model has developed to include the attack classification. The name of the model has also developed in tandem with the introduction of the attack classification in order that it be simpler to both (a) recall and (b) understand. The use of the word "dearth" in earlier work on the model was deemed unnecessarily confusing.

Purpose

The model is a broad classification model. It is proposed that it should be used to classify disasters in the first instance, with further classification likely required for further research and learning purposes. It is proposed that the model informs decision making with respect to resource and training prioritisation, for instance. It is not proposed that the model be absolute, and it is understood that the classifications are, in some respects, subjective.

24 The Disaster Database in Introduction to Crowd Science by Prof. Dr. G. Keith Still formed the basis of the disasters researched. More recent Crowd Disasters, occurring since the publication of Prof Still's book in 2014 have also been researched.

It is further proposed that the model will stimulate debate and potentially inspire a renewed focus among Crowd Safety professionals that there might be more work done in the area.

Assumption

It is assumed that the classifications Lack and Disregard are, in the most part, mutually exclusive. Broadly speaking, disasters either take place in an environment where there is regulation, knowledge and expertise to hand or where there is not. That being the case, the regulation, knowledge and expertise are either Lack(ing) or Disregard(ed). This is the basic assumption on which the model is developed.

The LaD Classification Model for Crowd Disasters

Lack

When establishing whether the Lack classification is appropriate, consideration is given to the presence or lack of:

- Relevant regulation/legislation
- Evidence of compliance-enforcement
- Evidence of general understanding and appreciation of crowd safety principles
- Evidence of police training specific to crowd safety and management
- Presence of competent crowd safety professionals operating in the jurisdiction/area
- Industry bodies driving development

It is possible through research to establish whether crowd safety is something embraced and taken seriously in a jurisdiction in which a crowd disaster takes place. Often, reports specific to a particular Crowd Disaster make it clear if a lack classification is appropriate. Beyond reports specific to a Disaster, there may well be a history of disasters in the area that may indicate the likely classification. It is no coincidence that certain jurisdictions experience disasters all too regularly.

Disregard

Crowd disasters are not confined to those jurisdictions mentioned above, where "lack" might be the most likely appropriate classification. Many Crowd Disasters take place in countries that have long understood the requirement for regulation and education in Crowd Safety. In establishing if Disregard is an appropriate classification, consideration is given to evidence of any disregard of:

- Relevant regulation/legislation
- Enforcement orders, permit denials, cease-and-desist orders, etc.
- Expert reports/consultancy projects specific to the event
- Available expertise and knowledge
- Established guidelines, be they national or widely adopted international guidelines

International guidelines and best-practice are key considerations within the LaD Model. Many countries and jurisdictions operate a "best-practice model" rather than relying on local regulation and legislation. In some cases, this is due to local regulation and legislation being outdated and needing updating while in other cases there are no local regulations or legislation with respect to Crowd Safety. This can be a key determining factor in assigning a Lack or Disregard classification.

The Republic of Ireland, for instance, has very little in the way of legislation and guidance with respect to events and crowd safety. The guidance that does exist is outdated and virtually inapplicable to modern day crowd safety.[25] Thus, those operating in the crowd safety field utilise key UK guidance such as The Green Guide, Safety at Sports Grounds, the Event Safety Guide, etc.

Attack

The attack classification is included because of the increase in attacks on places where crowd gather in recent times. Realistically, a well-planned and targeted attack on an event or crowded situation is extremely difficult to combat in advance. It is predominantly about having the plans and people in place to deal with it once it has happened.

While the focus should be on the plan to deal with the attack on the crowd once it has happened, there are ways the attacks might be avoided. Inspector Dan Forster, an experienced Police officer with Avon and Somerset Constabulary in the UK, with specific expertise in events recently noted that

> One of the key principles of crime prevention is target hardening. The most secure locations have layers of security measures that will be both physical and procedural. Many events already have layers of security in place, and in that sense, may be less attractive to attackers that other crowded areas, such as shopping centres and transport hubs[26].

[25] The Code of Practice for Safety at Outdoor Pop Concerts and Other Outdoor Musical Events (POP Code) and The Code of Practice for Safety at Sports Grounds were released by the Department of Education in 1996.

[26] http://www.eventmanagerblog.com/terrorism-and-events

It is important to not let a sense of inevitability pervade with respect to attacks on crowds and events. The increase in instances of same should focus the minds and efforts of those running events and managing crowds to ensure their plans and resources are robust enough to effectively deal with the aftermath of attacks.

It is proposed that more work will be done in this area within the industry, arising out of the recent proliferation of attacks.

Notes

Classifications made through use of the LaD Model are done with a view to informing learning and avoidance of similar crowd disasters in the future. They are not designed to be a determination of fault or guilt and are usually done sight-unseen of likely key documentation and information not available to the public.

Currently, a classification as an Attack within this model results from a first pass review of the circumstances of the disaster. It does not speak to whether, beyond that, a classification of Lack or Disregard might also be applicable.

It is proposed that the impact of the Attack classification and its usefulness in illustrating the recent significant upswing in occurrence of attacks on crowds is important and has potential to help bring about renewed focus and change within the industry. For now, the Model has not been developed beyond a single classification per disaster.

Examples of crowd disasters

These examples illustrate that crowd disasters are an ongoing issue and are not particular to any region or event-type. This sample is of relatively recent disasters, underlining the importance of crowd disasters as an area of focus and study. These disasters are happening now.[27]

Nice attack, France, 2016

Eighty-six people were killed on 14th July 2016 in Nice, France, when an attacker drove a 19-tonne lorry into a large crowd gathered on the promenade celebrating Bastille Day. The crowd was gathered to enjoy a firework display when, at approximately 22:30 local time, when the attacker struck, driving approximately 2km along the promenade through the crowd before police shot and killed him. The so-called Islamic State claimed responsibility.[28]

[27] This sample shows an even distribution of classifications, which is done for illustrative purposes, and should not be taken to be representative of any broader sampling or classification exercise.

[28] http://www.bbc.com/news/world-europe-36801671

Paris attacks, France, 2015

In a series of coordinated terrorist attacks in Paris on 13th November 2015, suicide bombers and gunmen armed with assault rifles killed over 130 people. Eighty-nine of those that died were killed at the Bataclan theatre during an Eagles of Death Metal concert.[29] The attackers were ultimately killed by police or blew themselves up. The group known as ISIS claimed responsibility for the attacks.[30]

Besiktas bombings, Istanbul 2016

The attack involving a car bomb and a suicide bomber took place on 10th December 2016, shortly after local team Besiktas finished a football match in the Vodafone Arena in Istanbul Turkey. Forty-six people died, 36 of those police officers, when first the car bomb was deployed outside the Vodafone Arena and shortly thereafter a suicide bomber attacked at the nearby Macka Park.[31,32]

Oakland fire, United States, 2016

Thirty-six people died when a fire broke out in a large warehouse venue being used to stage an electronic music show. The venue was inhabited and run by an artists' collective. This practice is common in the area.[33] While the exact cause of the fire is not yet known, investigators have identified the area in which the fire began as being at the rear of the warehouse, on the 1st floor.[34]

Kiss Nightclub Fire, Rio, Brazil, 2013

Two-hundred and forty-two people died when a fire broke out in the Kiss Nightclub in Rio, Brazil, on 27th January 2013. The fire was caused by a band's use of outdoor pyrotechnics inside the venue.[35] Two Brazilian firefighters have been convicted in relation to the fire, with one having been involved in a company that also did renovations on the venue.[36]

---

29 https://en.wikipedia.org/wiki/November_2015_Paris_attacks

30 http://edition.cnn.com/2015/12/08/europe/2015-paris-terror-attacks-fast-facts/

31 http://www.bbc.com/news/world-europe-38323326

32 https://en.wikipedia.org/wiki/December_2016_Istanbul_bombings#cite_note-cumhuriyet-1

33 http://www.newyorker.com/news/benjamin-wallace-wells/what-happened-exactly-in-oaklands-ghost-ship-fire

34 http://www.sfchronicle.com/bayarea/article/Oakland-warehouse-fire-ghost-ship-what-we-know-10796688.php

35 http://www.reuters.com/article/us-brazil-nightclub-idUSBRE90Q04T20130128

36 http://www.telegraph.co.uk/news/worldnews/southamerica/brazil/11650260/Brazil-court-convicts-two-firefighters-in-nightclub-fire.html

Hajj, Mina, 2015

Reports vary on the number of people killed and injured in the Disaster that happened on 24th September 2015 in Mina, Mecca, Saudi Arabia. Government reports have the death toll at 769 while independent estimates put the death toll at well over 2,000.[37] The Disaster occurred at the intersection of two heavily trafficked streets near the Jamarat Bridge. Hajj has a history of similar disasters.[38]

Shanghai, China, 2014

31 December 2014 saw 36 people killed during a Crowd Disaster at the New Year's Eve celebrations.[39] A reported crowd of 300,000 plus people had gathered to watch a firework show, which had already been cancelled by authorities. Investigations have laid the blame at the doors of officials and local government.[40]

Mumbai, India, 2014

On 14th January 2014, a reported 18 people died, and 56 people were injured while gathering to pay their last respects to a religious leader who had recently passed away. Reports indicate that the Disaster occurred when the gates to the religious leader's home were opened, and the crowd surged forward. Reports also indicate some people climbed gates to gain entry only to fall.[41]

Datia, India, 2013

A reported 91 people died with over 100 injured during the Hindu festival of Navrati when a rumour circulated that a bridge was about to collapse.[42] The rumour led to some people jumping from the bridge and others seeking to push their way off the bridge. Some reports claim a tractor hit the bridge, damaging it, while other reports point to a railing that was broken on the bridge as contributing to the Disaster.[43]

---

[37] https://en.wikipedia.org/wiki/2015_Mina_stampede#cite_note-Paul30Sept-81
[38] http://www.reuters.com/article/us-saudi-haj-disaster-idUSKCN0RU18R20150930
[39] https://en.wikipedia.org/wiki/2014_Shanghai_stampede
[40] http://edition.cnn.com/2015/01/21/world/china-shanghai-stampede/
[41] https://en.wikipedia.org/wiki/2014_Mumbai_stampede
[42] http://www.thehindu.com/news/national/other-states/91-die-as-rumours-spark-stampede-in-mp-temple/article5231189.ece
[43] https://en.wikipedia.org/wiki/2013_Madhya_Pradesh_stampede

Establishing the classification

A wide range of sources are researched by the user to inform the classification assigned to the Disasters within the Model. They include (Table 2.4):

- Media reports on the disasters
- Witness accounts of the disasters (publicly available)
- National regulation and legislation within the country of each disaster
- Any known expertise/professional operators within the country of each disaster
- History of Crowd Disasters in the country/jurisdiction

Table 2.4 LaD classification model of crowd disasters

Year	*Venue, location*	*Brief description*	*Classification*
2016	Istanbul, Turkey	46 people died, the majority being police officers, in an attack that saw a car bomb detonated as well as a suicide bomber attack.	**ATTACK**
2016	Nice, France	86 people killed when an attacker drove a truck into a crowd gathered celebrating Bastille Day.	**ATTACK**
2016	Oakland, California	36 people died during a fire in a warehouse that was being used to stage an electronic music show	**DISREGARD**
2015	Paris, France	130 people were killed in several coordinated attacks that included bombings and shootings.	**ATTACK**
2015	Hajj, Mina	Potentially over 2,000 people killed (reports vary) during a religious event with a history of similar incidents.	**DISREGARD**
2015	Shanghai, China	36 people killed during an NYE celebration that had been cancelled but saw 300,000 people show up.	**LACK**
2014	Mumbai, India	18 people killed during mourning of a religious leader who had passed away.	**LACK**
2013	Rio, Brazil	242 people died during a fire in a nightclub venue when outdoor pyrotechnics were used as part of an indoor show.	**DISREGARD**
2013	Datia, India	91 people dies when a rumour circulated that a bridge on which thousands of people were standing was facing imminent collapse.	**LACK**

- Peer-reviewed papers
- Reports from inquests, court cases, etc.
- Personal knowledge

No weighting system has been included in the Model. The research and application of the model has not shown any situations to date where a weighting would be helpful to determine the classification of a Disaster. The classification appropriate to each disaster has been quite clear. It is not proposed, at this stage, that weighting as a feature of the model be ruled out. This requires further research and investigation.

The importance of causality

The appropriate use of the model is predicated on the user's understanding of distal and proximate causality and the key differences between both. As noted by Still in Introduction to Crowd Science (2014, p. 76)

> A proximate cause is an event that is closest to, or immediately responsible for causing some observed result. This exists in contrast to a higher-level ultimate cause or distal cause that is usually thought of as the fundamental reason for an occurrence.

This understanding underpins the application of the Model, in that it ensures the user considers more than the obvious (proximate) cause of the Disaster. Taking the KISS Nightclub fire in Rio, Brazil in 2013 as an example, one might consider the proximate cause to be the use of outdoor pyrotechnics indoors, thus causing a fire. One might further conclude the distal cause to be the lack of proper implementation of regulations, which ultimately led to the jailing of two firefighters. Both are legitimate causes, but the proximate cause is of greater interest to us in the context of this Model.

It is important the user's research be thorough, allowing for proximate cause to be identified wherever appropriate.

Considering cognitive/conditional bias

Conditional bias has the potential to impact on the use of and efficacy of the model. It is important that users understand the concept of conditional bias, both in themselves as well as in media reports, witness testimonies, court affidavits, etc.

The users must be acutely aware of their own beliefs, preferences and leanings and must not seek, in applying the model, the classification (result) that best suits their own cognitive/conditional bias. For instance, my position that criminal negligence on the parts of multiple parties involved in the organisation and licensing of a particular disaster should not mean I seek to support that same conclusion in a different disaster exhibiting similarities. Human nature dictates that I am likely to do so unless I am acutely aware of my conditional bias and ensure that awareness comes to bear on my research and application of the model.

Cognitive/conditional bias is particularly evident in media reports as well as witness testimonies. The media uses terms like "crush" and "stampede" interchangeably when they should not. In the reporting of disasters, they often use emotive words in headlines and articles. These portray their own biases and can give rise to biases among those reading. In the case of users of this model researching media reports, witness testimonies, etc., a key task is to remove any apparent cognitive/conditional bias. Users must focus on the facts and establish if they support the apparent bias. They may well support it, but they often will not.

Conclusions

The model allows for straightforward classification of Crowd Disasters and is used to identify areas/countries/jurisdictions deficient in the presence of training/understanding/regulation/compliance-enforcement as well as areas/countries/jurisdictions where these exist but are disregarded. It is proposed that the model will aid in ultimately understanding where deficiencies leading to Crowd Disasters exist, thus allowing interested parties to seek to address those deficiencies with a view to improving Crowd Safety.

In the case of Attack classifications, it is proposed that the Model be useful as a central barometer of the increase in/proliferation of such attacks on crowds. This should be useful in graphing trends with a view to impressing upon various stakeholders how serious an issue attacks on crowds are becoming. This should be the first step in a global conversation and movement among Crowd Safety professionals to improve planning with respect to preventing and responding to attacks on our crowds.

Citation

The model has been purposely titled the LaD Classification Model for Crowd Disasters, with the "a" in LaD being lower case. This acts as a nod to the previously unpublished work by the author[44] on a classification model that included variations on the "L" and "D" but not the "a". It also sets the attack classification apart from the others. This is done for good reason, as outlined in this work.

44 Mark Breen, https://www.cuckoo.ie/our-team/

REFERENCES

Department of Education [ROI] (1996). 'Code of Practice for Safety at Outdoor Pop Concerts and Other Outdoor Musical Events.'

Department of Education [ROI] (1996). 'Code of Practice for Safety at Sports Grounds.'

Fruin, J. (1993). 'Cause and Prevention of Disasters', Elsevier Science Publishers B.B.

Still, K. (2000). 'Crowd Dynamics.' PhD Thesis presented to the University of Warwick, 2000.

Still, K. (2014). 'Introduction to Crowd Science', 1st ed., Florida: CRC Press.

DIM-ICE AND THE UK EVENT GUIDE TO SAFETY

We based the DIM-ICE risk model on the principle of *focussing* the users mind on the underlying causes of accidents and incidents, of design, information, and management failures. On how we can influence the crowds' expectations, and crowd safety, using the design, the information and management systems. Section 7 of the *Purple Guide to Health, Safety and Welfare at Music and Other Events*[45]states:

> CAPACITIES
>
> At every event, whether it is free or paid, outdoor or indoor, there is a need to establish a safe capacity. ... At events where those capacities are likely to be reached, it is important to establish a method for continuously counting those present to ensure it stays within the safe limits. ... Occupant capacities can be calculated for the available space for each area based on sight lines, topography, audience profiles and needs (e.g., transport hubs). The capacity of the venue may also be a condition on a license.
>
> CROWD DYNAMICS
>
> A great deal of research has been done on crowd dynamics and it is important that event organisers take this into account when planning. There are three main phases of crowd behaviour:
>
> - Ingress (e.g., access routes, queuing space and entrances).
> - Circulation (e.g., concourses, space around facilities).
> - Egress (e.g., exit routes and exit-gate widths).
>
> And three main influences on crowd behaviour:

[45] Which replaces the original "Purple Guide" (HSG195).

- Design.
- Information.
- Management.

> In considering crowd management, organisers need to consider likely flow rates and movements together with densities, particularly in key areas.

Our aim was not to dumb down any aspect of risk assessment or risk management. We do not intend to replace the existing systems. Our aim is to provide enhanced methods to understand crowd risk analysis as a dynamical system. The reason should be clear. When the crowd is in motion, the crowd dynamics risk is changing over time. You cannot reduce the description of the potential risks to single number and a brief line of description.

The DIM-ICE risk model is one of suite of analytical tools. It achieved our objectives, in that it is simple to understand, simple use, and simple to communicate crowd risks.

Chapter 3

Crowd dynamics

> The store of wisdom does not consist of hard coins which keep their shape as they pass from hand to hand; it consists of ideas and doctrines whose meanings change with the minds that entertain them.
>
> John Plamenatz

In this chapter, we outline the fundamentals of crowd density, crowd flow and the associated risks to the crowd in places of public assembly. We recap on the key points outlined and detailed in the *Introduction to Crowd Science*. We know that there will be repetition as we cannot assume everyone has a copy or has read the previous book. We also compare UK and USA guidance for crowd density in places of public assembly.

INTRODUCTION

The focus of our early work was the simulation of crowd dynamics, both the individual and the collective nature of people moving around places of public assembly. For years, we studied the movement of crowds in and around the Wembley Complex. The stadium, the conference centre, the arena and the two stations that served the area. We featured in a variety of science magazines and international TV programmes from 1996[1] on algorithmic development for a range of applications.

Developing fast algorithms for crowd simulations led to two inevitable conclusions. First, that all the rules of human behaviour may never be fully programmable, there were too many unknowns. Secondly, there was a predictable set of characteristics, an emergent behaviour that results from density, flow and geometry. In other words, if we know two of these (say geometry and density), we can predict the third (flow).

We also inverted the application of the simulations instead of trying to capture, calibrate and evaluate all the elements of human behaviour for assessing the safety of an area, we were tested how an environment might

[1] https://www.gkstill.com/News.html

DOI: 10.1201/9781351053068-3

fail? By this inversion, the simulations could provide insights to where and when a system might reach critical density and when we would need to intervene to keep an environment safe. For over a decade, we developed crowd dynamics algorithms and spatial analysis techniques using a range of mathematical approaches. It was a fun developing and exploring these models and we created a wide range of geometries to evaluate the wayfinding algorithms through built and complex spaces.

We explored how crowds form and move, when they form and move, and why they form and move. There is a natural overlap between the mathematics (how, when and where) and psychology (when, where and why). This was all interesting work, but in 1999, when the UK Cabinet Office Emergency Planning College first asked[2] the author to teach a course, it was apparent that there was a gulf between theory and application. For years, we had used simulations on major client projects and were, at that time, in the process of developing crowd models at the Sydney Olympics. Teaching the first class at the UK Cabinet Office Emergency Planning College showed us just how much work was needed to turn the theory into practical applications. Turn the simulations into processes, procedures and practical applications. In addition to understanding the crowd dynamics, we also had to develop a range of tools to *teach* the principles and applications of crowd dynamics.

THE LANGUAGE BARRIER

When teaching, we immediately ran into a series of significant issues. Initially, most crowd managers were not computer programmers or mathematically inclined. This immediately ruled out the universal application of computer simulations. Crowd managers were the people who made the crowd safety decisions at events, the approach (computing) was the wrong.

Then there was the cost problem. Crowd safety is an investment, like insurance. The impression is that you only need it if things go wrong. As such, it shows as a cost, a negative in the balance sheets. Staffing events used zero-hour contracts and volunteers. Budgets were minimal. Investing in computer simulations, which needed experts in creating, using, understanding and converting the outputs into practice, was not a route to implementing fundamental safety changes in the industry.

Another problem was the language used to describe crowds.

> Packed, congested, crushed, busy, cramped, full, filled, full to bursting, jam-packed, overfull, overflowing, teeming, swarming, thronged, populous, overpopulated, over-peopled, mobbed, wall-to-wall, stuffed, chock-a-block, chock-full, bursting at the seams, bulging at the seams,

[2] During a research visit to the library (1999), John Parkinson (librarian), showed me a file on my work. Steve Hick (Course Director) asked if I could teach a class, the following week.

packed to the gunwales, like Piccadilly Circus, chocker, mobbed, throng, a sea of people, overcrowded.

These are words used to describe crowded spaces. We heard the term "Crushed" used to describe both a "bit busy" and "the imminent risk of significant injury". We saw after incident reports that one eyewitness described as "normal" and another as "high-risk". This was a significant challenge because it communicates the risk of red-flagging problems, much more difficult. Confusing language and incorrect/inappropriate words lead to confusing communications and can lead to poor decision making[3].

Language can be highly subjective and emotive. It is essential to understand that the language you use, your vocabulary and the context in which you are making your observations need to be clear throughout. It would be best if you used a common language; otherwise, confusion, indecision and delays will inevitably result. A delayed decision can be the difference between reacting in a timely manner to an incident or reacting too late to do anything to prevent a situation from developing into an incident.

As we taught around the world, we also noted another problem. The word "safety" and the word "security" in different languages are the same. There is a difference between the fundamentals of a crowd safety plan and an event security plan, but this was a blurred division, and the language did not clearly differentiate between these two different concepts (Table 3.1).

A crowd safety plan and a crowd security plan are two different things. There may be some overlap in the ideas, but they are different concepts, where crowd safety should focus on the design, the information and the management (process/procedures) for the crowd. The security concept would relate to (say) counterterrorist, potential treats of attack and systems security (such as access passes). The challenge was to create a suitable language to define a safety concept, define the limitations in the system

Table 3.1 Safety and security in different languages

	Albanian	*Belarusian*	*Catalan*	*Croatian*	*Chinese*	*Danish*
Safety	Siguri	бяспекі	Seguretat	Sigurnosni	安全	Sikkerhed
Security	Siguri	бяспекі	Seguretat	Sigurnosni	安全	Sikkerhed

	Dutch	*French*	*Greek*	*German*	*Italian*	*Malay*
Safety	Veiligheid	Sécurité	Ασφάλεια	Sicherheit	Sicurezza	Keselamatan
Security	Veiligheid	Sécurité	Ασφάλεια	Sicherheit	Sicurezza	Keselamatan

	Nepal	*Swedish*	*Polish*	*Ukraine*	*Yiddish*	*Zulu*
Safety	सुरक्ष	Säkerhet	Bezpieczeństwo	Безпека	זיכערקייַט	Ukuphepha
Security	सुरक्ष	Säkerhet	Bezpieczeństwo	Безпека	זיכערקייַט	Ukuphepha

[3] We return to this topic in later chapters of this book.

design, highlight information requirements and aid crowd management strategies. We need a vocabulary specifically related to crowd safety and crowd dynamics.

THE GREEN GUIDE

The Guide to Safety at Sports Ground, called the Green Guide, is a UK publication. The first Guide to Safety at Sports Ground was published in 1973 followed, in 1986, by the second edition. The third edition appeared in 1989, with the fourth (1997) and fifth appeared in 2008. The current version, the sixth edition, appears in 2018.

Between fourth and fifth editions, the influences on crowd behaviour started to feature more prominently in the Green Guide. In the sixth, 2018 edition, there is a clear definition of crowd density in Section 9.

The sixth edition of the Guide to Safety at Sports Grounds, in section 9, states that

> An understanding of crowd density levels are vital for the safe management of concourses and circulation routes.

Which highlights the issue of the use of inaccurate language or descriptions. What do we mean by crowd density? Crushed, packed, crammed…?

VISUALISING CROWD DENSITY

We developed a software modelling tool; it does not have any simulation capabilities but does provide visuals of crowds (Figure 3.1).

In the classroom and the online courses, we ask the learners to write down the density they think the above (and other) images display. Numbers range from 1 to 10 people per square metre. Studies by MIT Neuroscientists have proven that the mind can identify images seen for as little as thirteen milliseconds[4]. However, without a frame of reference, a visual comparison, estimating density was highly subjective and often incorrect. This highlights the need for a visual identifier since there is a major difference between 1 (safe) and 10 (dangerous) crowd packing densities. Figure 3.2 shows the same crowd, from the top-down view, giving the viewer a distinct perspective of the crowds in the above images.

All the images we show are two people per square metre (1.7 people per square yard), and hence relatively[5] safe. A modelling tool, we provide as

[4] https://news.mit.edu/2014/in-the-blink-of-an-eye-0116

[5] We define these later in this chapter.

Figure 3.1 Computer-generated figures showing crowd density from different angles.

part of our *Level 5 – Introduction to Crowd Science*, online programme, allows the user to import a map, plan, or diagram, then define an area, then populate this with 3D characters. They can then shift the point of view (PoV) to match CCTV angles. This provides a visual aid for decision-making. Whether you are at ground level, with a high vantage point, or an ariel image, knowing the density, aids your decision-making. This tool highlights that, depending on your angle, your line of sight, your perspective, your point of view, you may perceive a high density (Figure 3.1) look higher density than Figure 3.2. Only the high angle, downward view, provides an idea of the space between the figures.

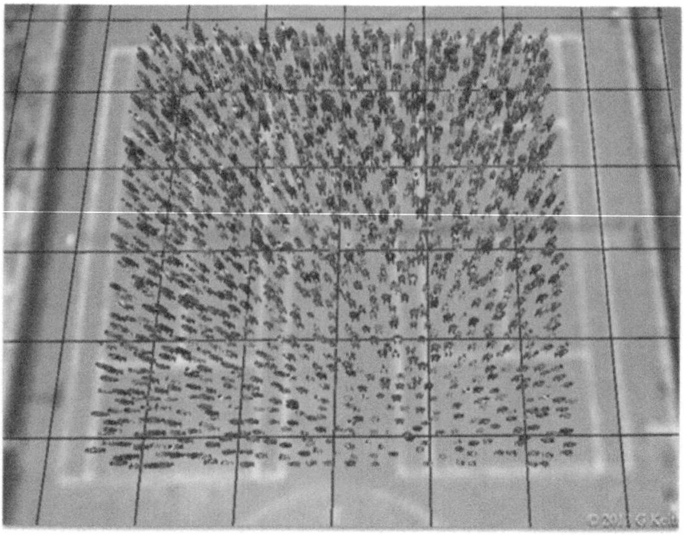

Figure 3.2 Two people per square metre (1.7 people per square yard).

DENSITY IS NOT ONE VALUE

Over the last fifty[6,7,8,9] years, the average British man has grown three inches taller, weighs 23 lbs more and has developed a paunch. Different environments may have different density limits. For example, an acceptable density at the front of the stage at a concert may not be acceptable in a high street. There are environments (concerts) which may have a higher population of younger, smaller framed, patrons. Other events may attract family groups, moving as one unit, with prams and pushchairs, hampers, folding chairs and other objects. Different demographics may have different body space[10] sizes with different nationalities having smaller, or taller, thinner, or more rotund than other nationalities.

It is important to understand that you should ***not*** apply any single figure (say 4.7 people per square metre) without considering the event, the crowd, their demographics and their spatial requirements. We can, of course, ***start*** with the density value, ***then*** we factor the people/profile of the crowd into the risk assessment.

6 https://en.wikipedia.org/wiki/Human_height
7 https://en.wikipedia.org/wiki/Human_body_weight
8 https://www.independent.co.uk/life-style/health-and-families/how-mens-bodies-have-changed-over-time-what-they-used-look-average-a7657881.html
9 https://www.mirror.co.uk/news/uk-news/uks-mr-average-extraordinary-change-10127526
10 https://www.academia.edu/23993416/BodySpace_Ebook_Ergonomic

You may calculate for one person per square metre/yard, but crowds do not evenly distribute across an event space at one person in each square metre/yard. People will cluster, typically in social groupings, there will be bubbles of space around clusters.

The clustering is essential to understand in the post-covid analysis of space usage. The assumption that you have provided space for social distancing, may fail because people will naturally want to cluster together, to talk and be social. You will have spaces between groups and higher density within groups.

One point to note is that when you observe "bubbles and clusters" you know the density is low. If the space starts to fill, an increase in crowd density, then we start to see, a more even distribution, the "bubbles and clusters" starts to merge. Figures 3.3 and 3.4 show the distribution change.

You can download a complete set of these images from the online courses. We have sets for 625m^2, 100 m^2, and 25 m^2. We also have a set of people per square feet (density) and square feet per person (Level of Service) for the USA learners. You can find all of these images from links on the course website.

https://crowdrisks.com/crowd-graphics.html

From the link above, you can print a set of images to aid your identification and classification of crowd density for planning an event and for

Figure 3.3 Density from one person per square metre to four people per square metre.

Figure 3.4 Five people per square metre.

identifying crowd density during an event. For example, for the 2011 Royal Wedding (William and Kate), we used the tool to plan the crowd management of the areas around Buckingham Palace. Knowing that there may be crowds arriving up to five days before the event, we can assign space, plan barriers, welfare, food and beverage concessions, toilets, security and other factors (Figure 3.5).

Figure 3.5 Royal Wedding (2011) planning image.

SIMPLIFYING THE VOCABULARY

The aim of these graphics is to replace the poorly defined vocabulary (*packed, congested, crushed, busy, cramped, full, filled to capacity, full to bursting, jam-packed, overfull, overflowing, teeming, swarming, thronged, populous, overpopulated, over-peopled, mobbed, wall-to-wall, stuffed, chock-a-block, chock-full, bursting at the seams, bulging at the seams, full to the gunwales, like Piccadilly Circus, chocker, mobbed, throng, a sea of people, overcrowded*) with a more clearly defined 1, 2, 3, 4, or 5 people per square metre (or the equivalent in square yards).

With correct visual definitions, we can now assign limits and risk definitions to these density values. We can use pictures to illustrate plans and reports. Using graphics in your plans means that everyone from planning through the licensing/permitting can understand the graphics in your event/crowd management plans. For example, the multi-agency groups working at an event, the first responders, the control room operators, even ground staff (print them, laminate them, provide a set to the stewards/ushers), to identify density and report accordingly are now talking the same language.

With a common vocabulary, we can now define density on a risk scale. Minimal risk (far left) at low density and substantial risk (far right) at higher densities (Figure 3.6). Again, this is the first pass analysis, you will need to factor demographics, events, spaces, physical and security, barriers, exits and furniture into your analysis.

We produced a tool, the 3D RTDS to assist in this process, with graphics for objects, picnics and characters of varied sizes.

RED, ORANGE (AMBER), GREEN

For decision-making, in real-time, in the control room, we can define 0%–50% of an area's capacity as Green. This would be 0–2 people per square metre. We can also define 50%–80% (3–4 people per square metre)

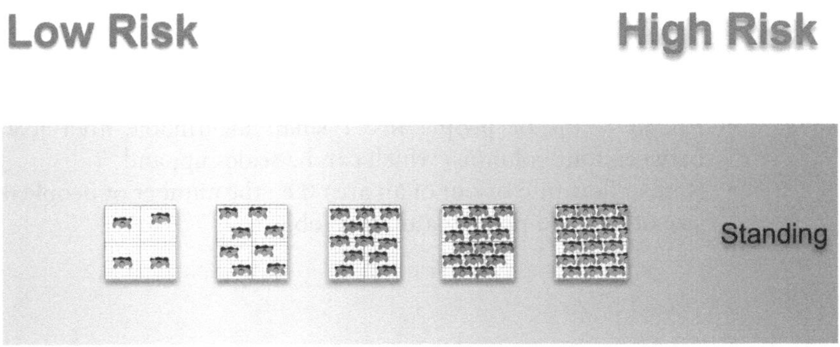

Figure 3.6 Green on the left, Orange/amber in the middle and red on the right.

as Orange (Amber) and Red as above 80% capacity (5 or more people per square metre). These are not arbitrary numbers, nor are they chosen at random. The UK Event guide uses two people per square metre as the initial capacity calculation. The British Standards (BS 8406:2020[11] Event stewarding – code of practice) outlines the following:

> 5.9 Escalation procedures
>
> There should be clearly defined procedures for management follow-up to incidents, and for response and support to stewards if incidents occur. If the steward does not contact the control point at agreed times, as specified in the assignment instructions, the supervisor should be notified and a visit to the static site made, or the relevant escalation procedure implemented.
>
> 5.10 Control point
>
> The organization should establish and operate a facility for control, co-ordination and monitoring of the event. Where practicable, the control point should be sited in a location that allows a view over the main site of the event and should facilitate lines of communication. The control point should monitor safety, including co-ordinating responses to incidents and emergencies.

However, this does not define the safe density or the definitions of risk that we outline in our courses and publications. The document costs £186.00 (at the time of writing this section) and is 16 pages long, mostly bullet points, coupled to brief descriptions. References to vocational training, outlined in the annex, offer additional training resources.

We can see the references point back to the Green Guide (the Guide to Safety at Sports Grounds) and Managing Crowds Safely[12]. The latter is a 92-page booklet which, on page 71, Section 220 states:

> Estimating the crowd
>
> 220. When monitoring the size of the crowd, the following are useful indicators
>
> - Space between people.
> - Rough count of people ins a small identifiable area (e.g.: between four columns) which can be scales up; and
> - Rate of flow into or out of an area (i.e.: the number of people or size of a crowd passing a marked object).

[11] https://shop.bsigroup.com/ProductDetail/?pid=000000000030393160
[12] https://www.hse.gov.uk/pubns/indg142.htm

Without a definition of safe density, these "estimates" are open to a wide interpretation. A review of the AUSSE course (*Level 2 Award in Understanding Stewarding at Spectator Events*) shows that there are no clear definitions of safe density, or how to communicate risks during an event. It does state that:

> There can be many serious and fatal consequences if an event is not well organised and if stewards do not carry out their roles and responsibilities in the correct way.

The AUSSE course definitions of crowd density states the following:
Overcrowding can be identified using simple methods

> Carefully observing and monitoring the crowd within your designated areas and if you can see the spectator's waist there is ample room in the area for more spectators.
> If you can see their shoulders the area is at capacity and no further spectators should be allowed to enter.
> If they can only see heads there is overcrowding and a strong possibility of crush injuries.

Not only is this definition of "density" both angle- and height-dependant, but also it does not provide a quantitative metric to assess safety or risk. You can see the point we are making here. If there are no clear definitions of how to assess risk or how to communicate those risks, how can we manage those risks?

When we first address this issue, back in 1999, we used a 1 metre square marked on the ground and physically put people in that square - demonstrating the density of 1, 2, 3, 4, 5 and more people per square metre. We discussed how this related to risk and demonstrated the onset of the shock wave, when people in the square are in physical contact[13]. We described that demonstration, and the moving density demonstration, in the previous book, *Introduction to Crowd Science*, and on our courses, we show videos and illustrations of this demonstration. The result was to produce a clearer, visual definition of crowd density risks, using graphics, from various angles, to facilitate easy identification from the ground (by stewards/ushers) to the control room CCTV angles and views.

4.7 OR 5?

The UK guidance states that you calculate the capacity of a standing areas based on your available space, using a maximum of 4.7 people per square metre. Then you apply the P and S Factors[14], exit capacity and a risk/threat

[13] Typically, this was around six–seven people per square metre but varied from country to country and class to class. The onset only starts when you have physical contact on all sides.

[14] https://sgsa.org.uk/p-and-s-factors/

assessment for a final capacity. Over the last few decades, we noticed the number of delegates who found it difficult to visualise the guidance of 4.7 people per square metre (or 47 people in 10 square metres) proved difficult. We needed a clearer method for spotting density in crowded spaces. In our classroom demonstrations we used a 1 square metre grid (or loop of string taped to the floor). For obvious reasons, it was not possible to use 4.7 people without increasing the area, and number of people, to a whole number. Hence the use of 5 people per square metre for visualising crowd density. We stress that this is a ***simplification*** to ***facilitate crowd spotting*** and ***identifying crowd density risks***, not for calculating your crowd density and crowd management plan for your event capacities. You must calculate your overall site capacity based on the guidance and legislation that are applicable at your specific locations.

USA – THE EVENT SAFETY GUIDE

When planning a major event, understanding crowd dynamics is an essential safety consideration. The USA Event Safety Guide (First Edition, Nov 21st, 2013) is a free-to-download document from the website[15] of the Event Safety Alliance. There is an outline on understanding the potential crowd and event risks, how to plan and manage safe events, and the guidance on relevant USA codes (e.g.: NFPA 101 – Life Safety code) in the USA Event Safety Guide. The text below is from the website of the Event Safety Alliance.

> … The Event Safety Guide compiles the best operational practices currently available in the live event industry in a single easily referenced manual.
>
> The guide is not a "how-to book" or a complicated set of standards. Rather, it is intended to help busy industry professionals know what safe workplace practices might be, heighten their understanding of the importance of safety in everything they do, and apply these best practices in their daily work.

This guidance contains safety considerations for crowded spaces, outlining a range of major events, the use of pyrotechnics, and is a collection of safety advice and best practices from around the world. Section 2, Page 9, of the Event Safety Guide states the following:

> 2.0.8 The event organiser is the individual or organization who promotes and manages an event. This role comes with many obligations and liabilities, and it is the organiser's burden to discover what those obligations are and the organiser's exposure to liability, keeping in

[15] https://www.eventsafetyalliance.org/the-event-safety-guide

mind those obligations and liabilities can change event to event, even if an event is replicated at a later date in the same location.

2.0.9 Organisers should consider obtaining legal and insurance advice early in the planning stage. Items that warrant consideration include:

- Liability for injuries.
- Liability for acts or omissions.
- Liability for financial obligations incurred in responding to major emergencies occasioned by the event, and
- Potential liability for the resultant effects of the event on normal emergency operations.

2.0.10 Planning an event can be complicated. Planning for the reasonably foreseeable risks and hazards associated with an event is even more difficult and essential to the event's success. Before scheduling the event, the organiser should consider the scope of the event, the risks to spectators and participants, community impact, and the emergency support required (personnel and logistics).

The event planning process needs to consider the health, safety and welfare of people attending the event. This includes members of the public and staff. Section 2 continues:

2.0.11 To protect the health, safety and welfare of people attending an event, as well as the event staff, contractors and subcontractors working at the event, health and safety has to be managed.
2.0.12 Some form of legislation usually governs or restricts public events or aspects of them. Some events, particularly extremely large or high-impact events, may require special state or local legislation.

Hiring a person or team, to navigate the workings of city hall, does not absolve the event organiser from potential labilities. You can "hire the best people" but will still be responsible for their output. Section also states that:

2.0.13 Organisers should assume the event will need a permit regardless of the size, location or timing. Investigate this inevitability early—many months before the event—to assure compliance. The permitting process is always intended to enhance safety and should be viewed as such.

When designing an event space, consideration for the safe occupancy limits (density), potential onsite crowd migration, and any elements in the design that could result in overcrowding and potential crowd risks. Evaluating the site occupancy (number of people attending, sometimes called the site capacity) starts with area analysis (the available space) minus the space used for infrastructure, stages, back of house, concessions, toilets, barriers, etc. Occupancy is not just the overall number of people on the site. It relates to the specific locations that a crowd may migrate towards, such as the front of stage areas, as the number of people per ft^2.

The USA Event Safety Guidance states the following:

> 8.4.1.1 Critical Crowd Densities (FEMA, 2010)
> 8.4.1.1.1 The objective should be to prevent large accumulations of patrons, particularly within short time periods, in confined spaces – especially if they cannot see what is happening.
> 8.4.1.1.2 A study by noted crowd behaviour expert Dr. John J. Fruin (1981) identified critical crowd densities as a common characteristic of crowd disasters. Critical crowd densities are approached when the floor space per standing person is reduced to about 5.38 ft^2 (0.5 square meters).

Dr. John J. Fruin is cited in several documents, they all use the same basic definitions for crowd density and crowd flow.

> Pedestrians require average areas of 24.73 ft^2 (2.297 square meters) per person to attain normal walking speed, and to pass and avoid others.
>
> At 10 ft^2 (0.929 square meters) per person, walking becomes significantly restricted, and speeds noticeably reduced.
>
> At 4.95 ft^2 (0.46 square meters) per person, the maximum capacity of a corridor or walkway is attained with movement at a shuffling gait and movement possible only as a group. This would be the characteristic of a group exiting a stadium or theatre.
>
> At less than 4.95 ft^2 (0.46 square meters) per person average, individual pedestrian mobility becomes increasingly restricted.
>
> At approximately 3 ft^2 (0.2787 square meters) per person, involuntary contact and brushing against others occurs. This is a behavioural threshold generally avoided by the public, except in crowded elevators and buses.
>
> Below 2 ft^2 (0.1858 square meters) per person, potentially dangerous crowd forces and psychological pressures begin to develop.

We note that there is no definition of the phrase "psychological pressure". Dangerous crowd forces below 2 square feet (0.1858 square meters) per person is clear, but there is a wide range of crowd phenomena, a spectrum of distresses from 24.73 square feet (2.297 square meters) per person down to the 2 square feet (0.1858 square meters). Proxemics are different for different nationalities. Proximity can feel threatening. From Proxemics – Cross-Cultural Concept of Space[16]

> Edward T. Hall, one of the founding fathers of intercultural communication, categorised the space surrounding people into personal space, social space and public space. Personal space refers to the 'bubble'

[16] https://www.communicaid.com/cross-cultural-training/blog/proxemics-concept-of-space/

surrounding a person which can only be entered by friends or close family members. In contrast, a person's public space is usually reserved for more impersonal and anonymous interactions. The layer of space between an individual's personal and public space is often called social space. This is the physical space where everyday contact takes place such as on a crowded train or bus.

Every culture has a set of hidden cultural rules concerning the physical space – or proxemics – between people when communicating. Breaking any of these rules could be interpreted as impolite or even threatening.

The USA Event Safety Guidance highlights the need for monitoring the crowd in places of public assembly.

8.4.1.1.4 Dr. Fruin (1981) wrote that "the combined pressure of massed pedestrians and shock wave effects that run through crowds at critical density levels produce forces which are impossible for individuals, even small groups of individuals, to resist".

8.4.1.1.5 The above information shows that you may need to provide a monitoring system, such as observers in strategic locations or closed-circuit television monitoring of crowd movements that will warn event personnel that they must take action to prevent a major incident.

NFPA 101 - LIFE SAFETY CODE

NFPA 101 – the life safety code in the USA has specific outlines relating to occupancy loading[17] (people per square foot). Despite its title, the standard is not a legal requirement, is not an instrument of law and has no statutory authority. However, some USA states adopt the code as their licensing and permitting codes, which can make the definitions contained therein a legal requirement. Table 3.2 is an extract from the *NFPA 101 Occupancy Load Factors*, with reference to event spaces.

Table 3.2 Extract from the NFPA 101 – occupancy load factors

Table 7.3.1.2 Occupant Load Factor (ft^2 per person)		
Use	(ft^2 per person)	(m^2 per person)
Assembly Use		
Concentrated use, without fixed seating	7 ft^2 per person	0.65 m^2 per person
Less concentrated use, without fixed seating	15 ft^2 per person	1.4 m^2 per person

[17] https://totalforcevlc.golearnportal.org/Fire_Inspector_II/Occupant_Load_Factor_Table.pdf

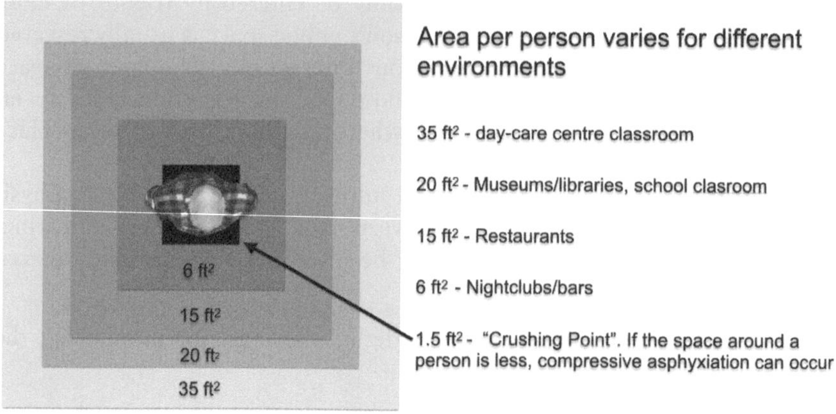

Figure 3.7 Illustration of area per person.

To illustrate crowd safety, Figure 3.7 shows the relative space around a person.

Which gives us a visual aid to understanding the area per person limits. NFPA 101, Life Safety Code, Section 12.1.7, states:

> 12.1.7.1 General. The occupant load, in number of persons for whom means of egress and other provisions are required, shall be determined on the basis of the occupant load factors of Table 7.3.1.2 that are characteristic of the use of the space or shall be determined as the maximum probable population of the space under consideration, whichever is greater.
> 12.1.7.1.2 In areas in excess of 10,000 ft² (930 m²), the occupant load shall not exceed one person in 7 ft² (0.65 m²).
> 12.1.7.4 Outdoor Facilities. In outdoor facilities, where approved by the authority having jurisdiction, the number of occupants who are each provided with not less than 15 ft² (1.4m²) of lawn surface shall be permitted to be excluded from the maximum occupant load of 6000 of 12.1.7.3 in determining the need for a life safety evaluation.

Section 8.4.1.3 of the Event Safety Guide states

> 8.4.1.3 The venue capacity depends upon the available space for people and the number of emergency exits. The latter is the subject of a calculation involving the appropriate evacuation rate, i.e., width of available exit space and appropriate evacuation route.

The emergency exit calculation is based on the localised occupancy and the flow rate (capacity) of the gate. From Section 8.4.1.3 of the Event Safety Guide states

> 8.4.1.7 Once you have a proposed capacity figure, meet with the local authority with jurisdiction (typically the fire department) over the determination of the venue capacity and review your numbers. During this

> meeting you will also need to present your exiting plans along with dimensions of those exits.
>
> 8.4.1.8 It is essential for effective crowd management during emergencies and evacuations that a well-planned and efficient means of escape exists for all occupants of the venue. ...
>
> Calculate exit door flow rates at 40 persons per minute for a single, fully open door (36 inches or 0.9 m wide) and 60 persons per minute for wider passages.
>
> Use 24 inches (0.610 m) when calculating the width of a person over the width of wider passageways. For example, if a passageway is 24 feet wide, 24 inches goes into 24 feet 12 times. Therefore, 12 is the maximum estimated number of people that can fit down that passageway at a time. Multiply the 12 persons wide by the 60 persons per minute mentioned above, and the result is a maximum exit capacity of 720 persons per minute.
>
> Note: This formula is intended only to acquaint one with the basics of egress space estimation and is not an official means by which egress capacity is measured. .... work with the local fire and building authorities having jurisdiction to establish exactly what is required in terms of means of egress.

The National Fire Protection Agency (NFPA[18]) sets the codes (standards) for crowd safety in places of public assembly. Both the event Safety Guide and NFPA 101 Life Safety Code highlight the need for trained crowd managers to monitor and manage crowd safety in places of public assembly. NFPA 101, 12.7.6, states that:

> ... in assembly occupancies having occupant loads exceeding 250, trained crowd managers or crowd manager supervisors shall be provided at a ratio of 1 crowd manager/supervisor for every 250 occupants.

The references that relate to crowd management in the NFPA 101 Section 12.7.6 define the requirement for crowd managers as follows:

> 12.7.6.1 Assembly occupancies shall be provided with a minimum of one trained crowd manager or crowd manager supervisor. Where the occupant load exceeds 250, additional trained crowd managers or crowd manager supervisors shall be provided at a ratio of one crowd manager or crowd manager supervisor for every 250 occupants, unless otherwise permitted by one of the following:

There are references to training for crowd managers in the NFPA 101 Life Safety Code

> 12.7.6.2 The crowd manager and crowd manager supervisor shall receive approved training in crowd management techniques.

[18] https://www.nfpa.org

> 12.7.6.3 Duties and responsibilities for the crowd manager and crowd manager supervisor shall be documented within a written emergency plan as required by 12.7.13.

CROWD SURGE

One of the hazards[19] in high density crowds is the phenomenon of crowd surge and crowd sway. These can be a significant risk to the safety of the patrons in high density crowds due to surge/sway. From the Event Safety Guide, Section 9.9 we note a specific reference to crowd sway:

> 9.9.2 Think carefully about where to position security and event personnel to monitor the audience for distress, crushing, sway, surges, or mosh pits as they all present a risk to the audience. Use of CCTV and/or the provision of raised viewing platforms, especially stage left and stage right, may help monitor the audience for signs of distress.

The USA guidance highlights that trained crowd managers are essential to plan, manage and monitor the crowds during the event. Trained crowd managers should be able to both understand and identify crowd risks. Provision of one trained crowd manager per 250 patrons is a requirement; depending on local state legislation, this might be a legal requirement.

CROWD RISKS

The Event Safety Guide outlines the parameters for crowd density and crowd flow rates through emergency exits. The reason this is so important is to both understand these crowd risks for events, to evaluate the safe evacuation times, and to avoid crushing at both entry and exit points, under both normal and emergency conditions.

Fruin[20] describes the forces involved in crowd related incidents.

> Crowd forces can reach levels that (are) almost impossible to resist or control. (...) Evidence of bent steel railings after several fatal crowd incidents show that forces of more than 4,500N (1,000 lbs) occurred. Forces are due to pushing, and the domino effect of people leaning against each other.

---

[19] **Risk** is the chance or probability of a person being harmed or experiencing an adverse health effect if exposed to a **hazard**. Example: A wet floor is a **hazard**, and there is a probability (**risk**) that someone might be harmed by slipping and falling.

[20] The causes and prevention of crowd disasters by John J. Fruin, Ph.D., P.E. United States of America (Originally presented at the First International Conference on Engineering for Crowd Safety, London, England, March 1993. Elsevier Science Publishers B.B. © 1993).

> Experiments to determine concentrated forces on guardrails due to leaning and pushing have shown that force of 30% to 75% of participant weight can occur.
>
> In a US National Bureau of Standards study of guardrails, three persons exerted a leaning force of 792N (178 lbs) and 609N (137 lbs) pushing.
>
> In a similar Australian Building Technology Centre study, three persons in a combined leaning and pushing posture developed a force of 1,370N (306 lbs).
>
> At around 3,000N (from medical bone fracture analysis[21]) there is a potential for a 5-person deep 'push' to break ribs....
>
> Gross Bone Fractures (Pathological and physical parameters[22]) are noted at forces around 6,920N...

From the same paper under the title "Prevention of crowd disasters by crowd management" Fruin states:

> Most major crowd disasters can be prevented by simple crowd management strategies. The primary crowd management objectives are the avoidance of critical crowd densities.

Understanding crowd density is an essential element of the overall crowd safety plan. Understanding how quickly and easily accidents and incidents can occur in crowds should be part of the overall safety considerations for planning for and managing crowds in places of public assembly.

From Fruin, if the area per person is less than 1.5 square feet, then the crowd density will reach the "crushing point". When exposed to high density, individuals are at significant risk to life and limb, to general surging and swaying in the crowd and crowd pressure waves.

Prior to any event, assess the risks of localised overcrowding (such as the front of stage barriers and areas where the geometry might focus crowds, inter-site routes, etc.), to prevent dangerous crowd build and excessive crowd density.

Monitoring crowd density across the site should be part of the event safety and crowd management plan, coupled with a risk management strategy to monitor and prevent dangerous overcrowding. Trained crowd managers should be monitoring crowd density. Events should designate a trained crowd manager to watch for (crowd spotters), and warn of, high density.

The use of visual aids to identify crowd build-up and high density, using the scale of 1–5 people per square yard/metre, can simplify the risk assessment and risk reporting.

---

21 From "Biomechanics of Chest and Abdomen Impact" (David C. Viano, Albert I King) 2000 CRC Press LLC.

22 From "Patellofemoral Joint Fracture Load Prediction using Physical and Pathological Parameters" Proceedings of the 1998 SAE International Congress and Exposition. SAE 980358 by PJ Atkinson, CM Mackenzie and RC Haunt.

HIGH DENSITY

As high density can build up around the front of stage areas, you may need pressure barriers, a pit crew trained to identify high density crowd risks and rescue distressed patrons, CCTV (to monitoring the crowd), and a clear protocol for reporting crowd problems.

This should be all included in the event safety plan[23]. For crowd safety at events, it is also important to understand how the performer may influence the crowd behaviour. For example, the style of the performance, if there were any reports of previous crowd safety issues related to the headline acts. Event risk assessment should consider the event, its history, the performers, and the nature of the crowd attending.

UK AND USA GUIDANCE

The USA and UK guidance documents require both study and understanding to fully appreciate the dimensions of crowd safety. For example, in the USA, the use of *area per person* is difficult to visualise, although this is the inverse[24] of density.

Therefore, we use graphics in our training programmes, giving the user simply to understand visual aids for their crowd safety management. However, there are fundamental differences between USA and UK guidance for crowd safety and the limits of crowd safety based on the physics of crowded spaces. First, we need to explore these, firstly for density, then flow, to fully understand the difference between guidance, licensing/permitting requirement, and physical safety.

VISUALISING CROWD DENSITY

The following graphics illustrate a body space, and you can download and use the set of density graphics from https://crowdrisks.com/crowd-graphics.html, which includes both metric and imperial (USA) versions.

In Figure 3.8, we show the average body size of both a male and female in a square yard and a square metre. Using the graphics from the 3D RTDS tool, we illustrate 1, 2, 3, 4 and 5 people per square yard/metre below (Figures 3.9–3.11).

---

[23] "Spotters should report high density" without a definition of high density can be open to interpretation.

[24] Density is equal to the reciprocal of area per person and vice versa (density = 1/area per person and area per person = 1/density).

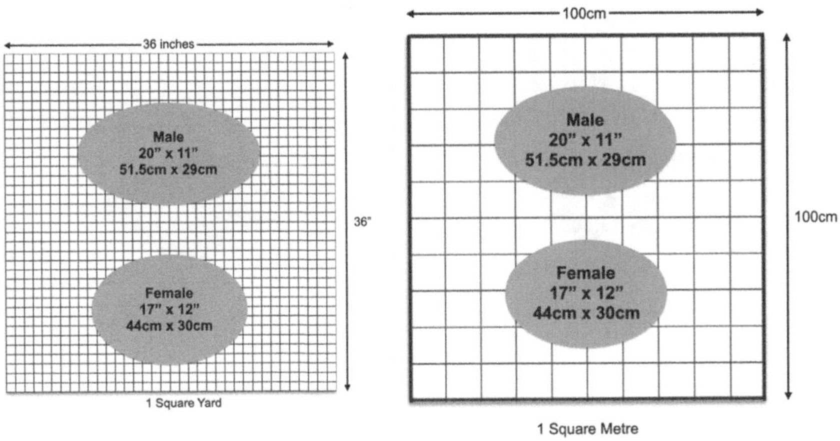

Figure 3.8 Average personal space (yards and metres).

LEGISLATION AND GUIDANCE

The UK Events Guidance (the Purple Guide) states in the introduction[25]

> The guide aims to help those who organise music or similar events, so that the events can run safely. As an employer, the event organiser - whether an individual, collective or local authority - has a general duty to ensure, so far as reasonably practicable, the health, safety, and welfare of their employees.
>
> ... At all stages, those using this guidance are encouraged to undertake appropriate risk assessments and evaluations to assess the specific requirements for each event they are organising or involved with. This guidance is not the law and there may well be other equally effective ways of dealing with risk.

The guidance, coupled with the legislation, covers the fundamentals of the duty of care of the employers, the organisers – whether individual, collective, or local authority – to the safety of their staff and patrons.

SAFE CAPACITIES

There are several considerations required to evaluate safe capacities for the UK Events. Previous versions of the Purple Guide stated that two people per square metre (0.5 square metre per person, two people per square metre).

[25] https://www.thepurpleguide.co.uk/index.php/the-purple-guide/78-introduction

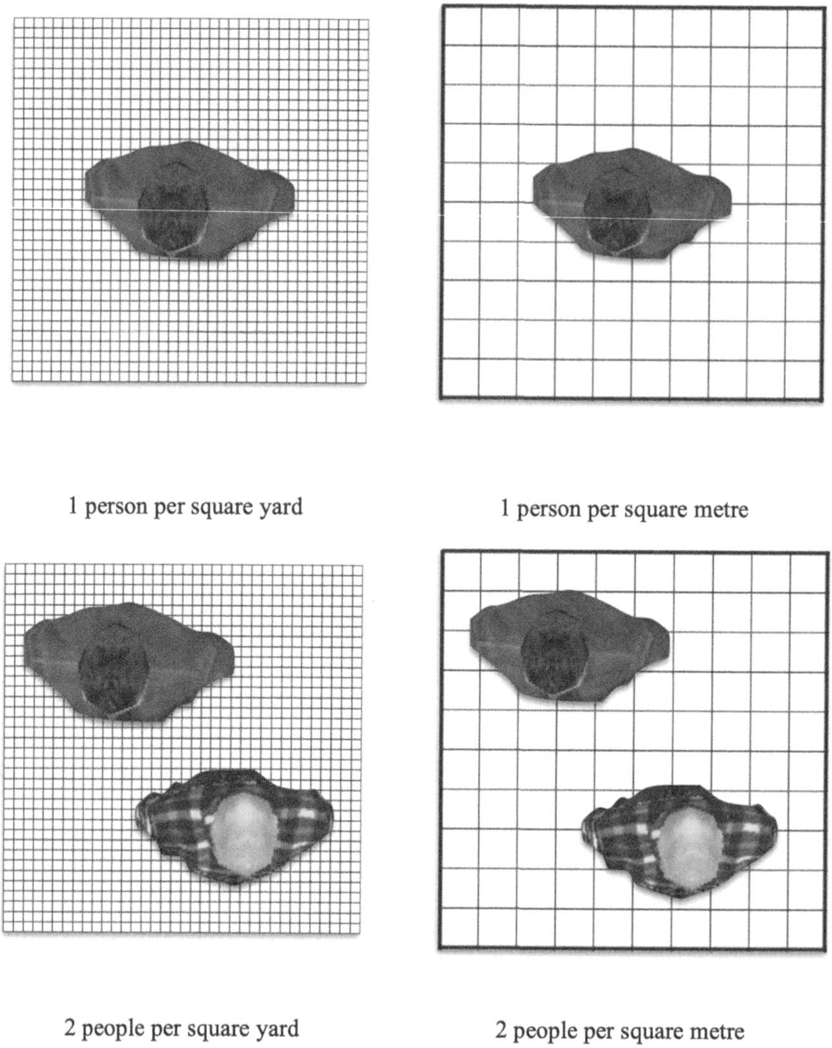

Figure 3.9 One to two people per square yard/metre.

> In venues where seating is provided, a major part of the occupant capacity will be the lesser of the two figures determined by the number of seats and exit provision. However, in other cases a calculation based on the acceptable occupant density should be carried out. Generally, 0.5m^2 of available floor space per person is used for outdoor music events.

From the UK Guide to Safety at Sports Grounds (6th Edition – Figure 11) we note the following density (and area/space per person) (Table 3.3).

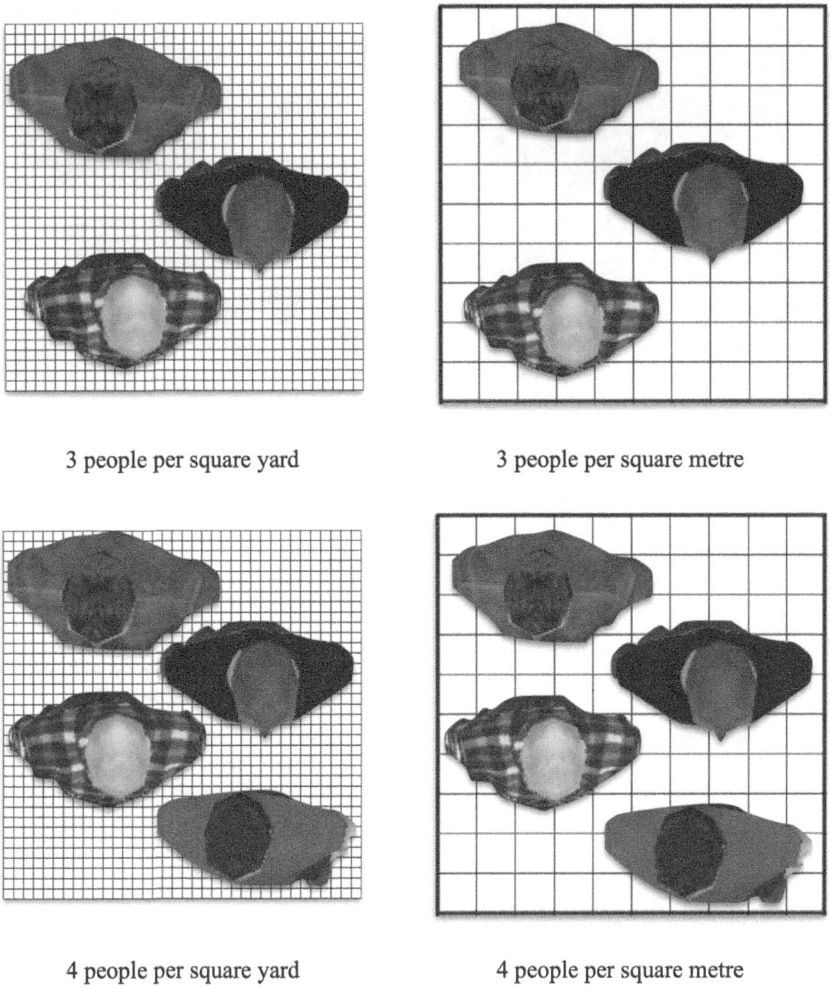

Figure 3.10 Three to four people per square yard/metre.

You can find similar density references on various guidance/planning[26] websites. The UK Green Guide states the upper limit of 4.7 persons per square metre[27] (0.21 square metre per person). This is the value you should use for static/standing crowds in your ***initial*** capacity calculations.

However, the total standing spectator area, in square metres, multiplied by 4.7, is only the START of your safety calculations. You also need to

26 https://www.designingbuildings.co.uk/wiki/Occupant_capacity_of_a_building_or_space

27 Note: this is the average sized person, which includes smaller profile body frames, family groups, larger body frames. Although the data shows the body sizes increasing over the last 20 years, and the average body size has increased, the values remain the same.

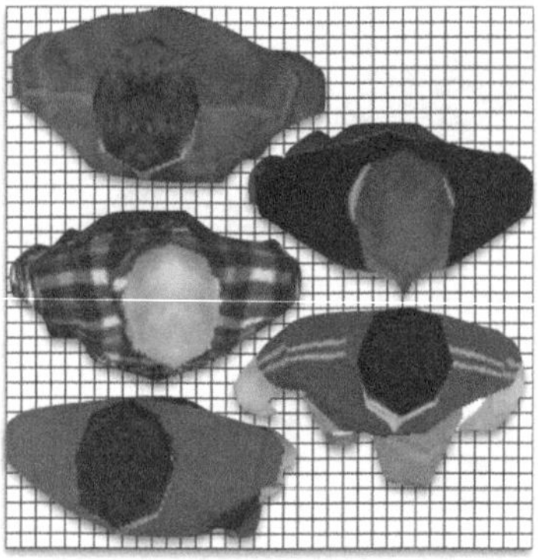

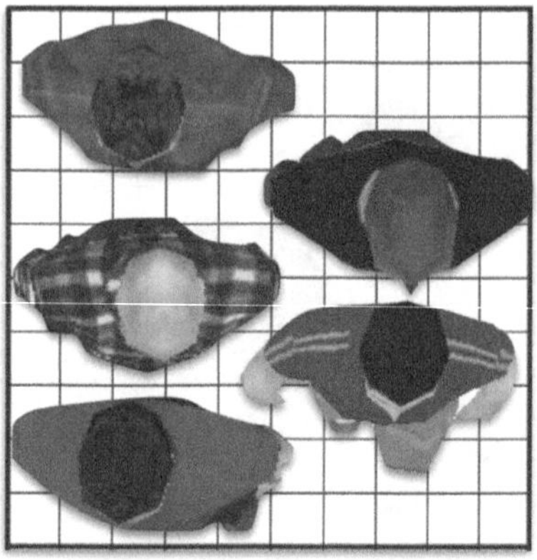

Figure 3.11 Five people per square yard/metre.

Table 3.3 Data showing crowd densities for stadia

Density per m²	*Density per 10 m²*	*Space per person (m²)*	*Details*
4.7 people	47 people	0.21	Maximum permissible for standing spectators
4.0 people	40 people	0.25	Preferred maximum density for standing spectators, and people queueing at bars/counters/toilets
3.3 people	33 people	0.3	Comfortable density for standing spectators/concourses refreshment areas.
2.0 people	20 people	0.5	Optimum density for general concourse areas (new construction) and for standing spectators on level areas at concerts
1.4 persons	14 people	0.71	More comfortable density for concourse/hospitality areas
0.7 persons	7 people	1.43	Optimum density for club style hospitality areas

assess the conditions of the site, the means of egress, the risks in the environment (pandemic[28] guidance) and the safety assessment. The UK Guidance outlines the use of P (physical) and S (Safety management) factors[29] cal-

[28] https://www.gov.uk/government/publications/how-to-stop-the-spread-of-coronavirus-covid-19

[29] https://sgsa.org.uk/p-and-s-factors/

culations. In other words, the area times 4.7 is the first pass approximate calculation for a maximum capacity of a spectator area at a sports stadium, ***not*** the final capacity figure.

COMPARING UK AND USA DENSITY/ AREA PER PERSON

The UK and the USA guidance differ in UK (metres) and USA (feet). If we compare UK and USA guidance data, we note:

> **UK** – maximum density permissible is 4.7 people per square metre, an area of 0.21 m^2 per person (2.26 ft^2 per person=1' 6" x 1' 6"=18" x 18").
> **USA** – in areas more than 10,000 ft^2 (930 m^2), the occupant load **shall not exceed** one person in 7 ft^2 (0.65 m^2). 7 ft^2 is 1,008 square inches, or 31¾" by 31¾"

Neither UK nor USA density evaluations factor the differences between male/female body sizes, larger or smaller body frames, family groups, children, elderly, and other factors. If we use the assumption of an "average" body frame (shown in Figure 3.8) then area and occupancy load are the basis of your capacity calculations.

UK AND USA DIFFERENCES

We can see there is a difference for site capacity between UK and USA. The local guidance, legislation, the safety licence (or permit) for the event will outline the *maximum* capacity for an event site. However, it is the ***localised areas*** of high density, and an early warning of a potential problem, that we need to address, the identification and reporting of high-density spaces and high-risk environments.

SPOTTING RISK

When crowds migrate across a site, say towards the front of stage, exit point, concession or move from stage to stage within a site, the overall site capacity does not change. However, the risks of high-density trip, slips and falls may increase as local spaces become more crowded. Whether we are working in feet/yards or metres, the risks increase when the density increases. The scales 1–5 people per yard/metre give us a visual indicator of those risks.

Using the average body frames, we can define 1–2 people per square yard/ metre as green. 3–4 people per square yard/meter as Orange/Amber and 5 or more people per square yard/metre as red for spaces that you would expect a stationary, spectating area, such as a parade, sporting event etc. We can now use this vocabulary, 1, 2, 3, 4 and 5 people per square yard/metre, Green, Orange/Amber and Red and our management decisions are 0%–50% (green) at 50% (on the cusp of orange/amber) get ready, and at 80% (on the cusp or Red) do not allow more people into the area.

The reason for making the "close site" at 80% will depend on the area. For the Royal Wedding (2011) we knew the capacity of the transport system and the distance between the underground/metro and the site. We had a good estimate of people in transit and needed a point to change the VMS (variable messaging system) to divert the crowd to Hyde Park, the overflow area. If we made that decision at 100% (full, >4.7 people per square metre as per the guidance) there may still be a crowd approaching the site. So, lock down at 80%, assess the situation and you can always allow more people into the site. But once it is full, once it reaches 100%, it will be difficult to turn people away.

By messaging, we can manage the crowds' expectations. By messaging "site full" before reaching 100%, we can influence their behaviour and divert them to an alternative site. For the Royal Wedding we added the subtlety of using the same message "to the Royal Wedding" and shifting the direction of the arrows. If the message read "To the Mall[30]" then shifted to "To Hyde Park" you can see how that can influence the crowd's decision. The people will not want to go to Hyde Park knowing the event is on the Mall. By labelling the signage as "Royal Wedding" we can divert the crowd to the safest areas to view the event and there were three massive screens at Hyde Park, to facilitate the crowd's view. In this, and other similar types of events, always consider the information, and how that will shape the crowds' expectations and behaviour.

OCCUPANCY

If you search on occupancy[31] or floor space factors, you will find there are density/floor space outlines for a wide range of environments. As we stated, earlier in this chapter, you may allocate space based on capacity and/or occupancy, but people may still cluster together. When we review places of public assembly, we would define sections at different, acceptable, densities. Pre and post pandemic, you can see how this helps you defined a safe, numerical, limit using graphics to determine when you reach that limit. Again, we stress the cut-off point at 80% of the capacity/occupancy. If you wait until you reach 100%, you do not have any margin for error.

[30] The Mall is the long road leading from Buckingham Palace to Admiralty Arch.

[31] https://www.designingbuildings.co.uk/wiki/Occupant_capacity_of_a_building_or_space

MOVING/DYNAMIC CROWDS

When the crowd is stationary, the maximum recommended density is 4.7 people per square metre, or 47 people in 10 square metres. In the classroom, we would stop there and ask the learners to sketch a shape with 10 square metres. It may not come as a surprise that many draw a square 10 metres×10 metres. That is, of course 1,000 square metres (10×10=100). It is a common mistake, and one that can lead to significant challenges in event planning.

We once noted a "senior" event safety consultant look at a map, during a site evaluation and comment that the 25 metres by 25 metres grid were 25 square metre sections, then work out its maximum capacity was 50 people[32,33], based on the 2 people per square metre event guidance in the UK. The actual area is 625 square metres; the capacity (using the event guidance of 2 people per square metre) is 1,250 people. We often see this type of gross error, in site calculations.

Which is why we emphasise splitting your site into triangles (1/2 base×height=area) or Google Earth Pro area assessment in the RAMP analysis.

A MOVING CROWD REQUIRES MORE SPACE

When a person extends their legs and arms, to walk, they occupy a larger area. Figure 3.17 illustrates this from a top-down view, shown in one square yard and 1 square metre. Arms and legs extend and swing forwards and backward when there is space to do so However, as we decrease the space around a person (increase the density), then movement becomes restricted (Figure 3.12).

In our Pre-Covid workshops, we used a 1 square yard/metre circle of string, described in *Introduction to Crowd Science*, to artificially force density on the crowd (by putting 1, 2, 3, 4, 5 people inside the circle). Videos of this experiment are on these links.

https://www.gkstill.com/Support/crowd-flow/2People.html
https://www.gkstill.com/Support/crowd-flow/3People.html
https://www.gkstill.com/Support/crowd-flow/4People.html
https://www.gkstill.com/Support/crowd-flow/5People.html
https://www.gkstill.com/Support/crowd-flow/5PeopleStep.html
https://www.gkstill.com/Support/crowd-flow/6People.html

We filmed those videos in Germany, in 2010, as part of a course. We have seen these clips appear, without accreditation, on other courses and websites.

[32] https://www.laeog.org/28.10.12/LAEOG_Density_V1.0.pdf
[33] https://www.stockton.gov.uk/media/1586058/9-crowd-dynamics.pdf

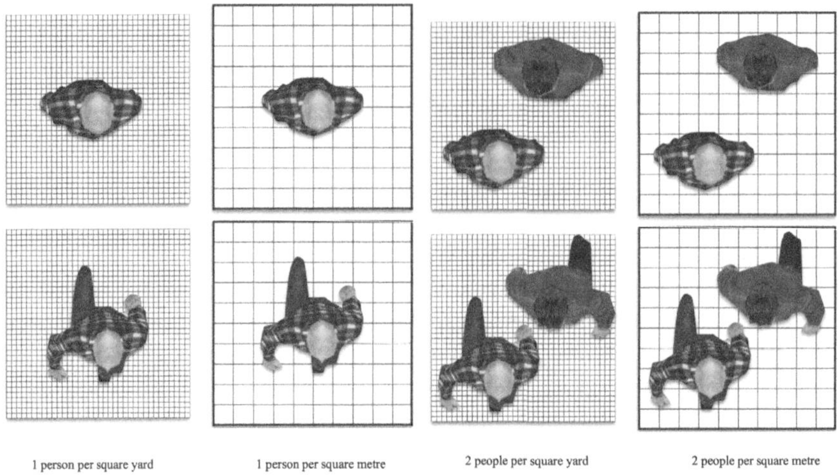

Figure 3.12 One and two people per square yard/metre – standing and walking.

We measured the time it took from this group to move 10 metres, and then plotted the results. The graph (Figure 3.13) is indicative of the relationship between crowd density (people per square metre) and crowd flow (speed of group movement).

If we compare the people per metre per minute and people per foot per minute, then 82 people per metre per minute=25 people per foot per

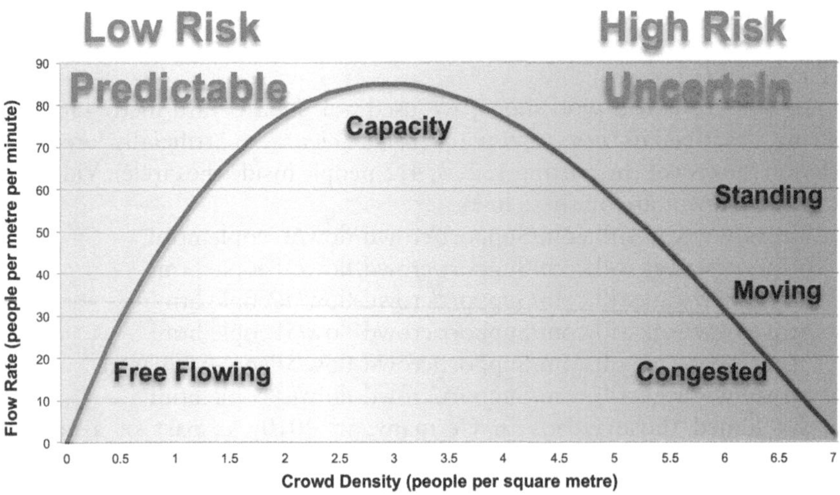

Figure 3.13 As density increases, the crowd flow increases, then decreases again.

minute[34] or 75 people per yard per minute. The curve for density against flow is the same; the bottom scale is in whole people per unit of area (yards or metres).

VOCABULARY

We can see that as density increases, the flow rate (number of people passing a line, expressed as people per metre per minute) increases, reaches a maximum value, then drops. We call the peak the capacity (maximum) flow. At lower densities, the crowd is free flowing, as there is sufficient space for any individual to take complete, unimpeded steps. Above 2–3 people per square metre, the flow rate becomes congested. By words "free flowing" at "capacity" and "congested" and the associated "minimal risk" (hence predictable), we refer to a congested or "substantial risk" crowd, meaning that the crowd's density is greater than five, and any trip, slip, or fall could result in significant risk to life and limb.

This is, of course, a simplification, but the aim was not to produce a calibrated scale. The objective was to produce a ***vocabulary*** for reporting crowded spaces.

We can now use colours and words, together, to define the status of both static and dynamic crowds. The scales, yards/metres, are dependent on local units of measurement.

ORDERLY CROWDS

In certain environments, such as commuter crowds moving during rush hour on a busy station platform can display high flow rates. This is a homogenous crowd, all moving with the same pace/stride, like an army on the march. High density and high flow rates are not sustainable as it only takes on individual, out of step, to suddenly congest those kinds of environments. European standards, BS EN 13200-1:2019[35], and the USA Highways Capacity Manual[36] define the maximum flow rate of 82 people per metre per minute, or 25 people per foot per minute (USA). This is the "capacity" part of the curve. In other words, the best you can hope to achieve is, at no more than three people per square metre, 82 people per metre per minute (or 25 people per foot per minute). When calculating the flow rate, you start with these values (optimal, best case), then factor into this the complexity

[34] 1 metre = 3.28 feet. 82 people per metre per minute / 3.28 (metres to feet) = 25 people per foot per minute.

[35] https://shop.bsigroup.com/ProductDetail/?pid=000000000030353511

[36] http://www.trb.org/Main/Blurbs/175169.aspx

of the route, the demographics, the weather, terrain, and other influences on walking speeds. These are the factors that will reduce the optimal, the maximum flow rate.

IF IN DOUBT, MEASURE IT

The easiest and most reliable method of flow rate evaluation is to measure it. Simply measure a section along a route, typically that might be your narrowest point, then count the number of people passing that section in a minute. If you measure an area and mark it on the ground, then you can take photographs of the number of people in the area (density) and the flow rate. Remember, if this is more than 1 metre wide (say 10), you will need to divide the number of people that pass the line, with the width of the section you are counting. In summary, if a route has the narrowest width of (say) 10 metres, then the capacity flow rate will be 820 people per minute. If density is lower, you will have a lower flow rate (not enough people passing the line). If the density is more than three people the flow rate will drop, due to the constraints on individuals taking whole paces, i.e., congested flow. If you look back to the graph of density against the crowd flow, you can see for any given density. We can estimate the flow rate, and vice versa. In Figure 3.19, a density of five people per square metre, you might expect a flow rate of 58 people per metre per minute. A density of one person per square metre, a flow rate of around 50 people per metre per minute, might be expected. Please note that these are not calibrated values. It is an approximation, but it can indicate how long it might take to fill or empty areas. As the UK/USA curves on the graphs are the same, the same principles apply to both UK and USA.

Replace the density and flow rate with the relevant yards/metres, but the principles, and the vocabulary, can remain the same (Figure 3.14).

SIMPLIFYING COMMUNICATIONS

By using the same colour scales for occupancy, 0%–50% is green, at 50%–80% is orange/amber, and 80% or more is red. We have a three-colour scheme to defined risk, congestion, flow, occupancy and risk. We have simplified the communication process. What is the status of this crowd, green, orange/amber or red? You can see how that simplifies the vocabulary and the language used in the control room for decision making. Hence this allows you to understand the risk; how it is developing/changing over time. Therefore, if there is a requirement for an early intervention, everyone is using the same point of reference.

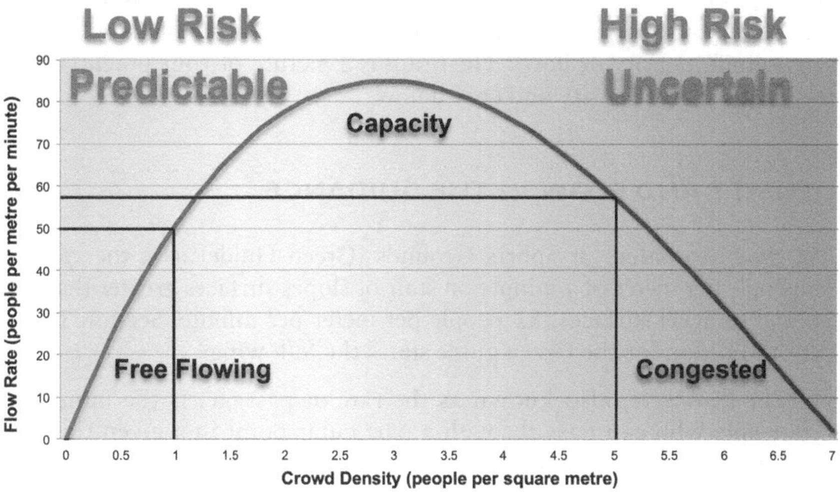

Figure 3.14 Flow rate (vertical) against density (horizontal).

You can think of this like a vehicle tachometer[37], red indicates risk. You can run a car engine in the red, but the likelihood of damage increases the longer you run the engine in the red. To manage risk, you need to keep your engine operating in the green or orange/amber. This does not mean a system will fail if it drifts into the red, only that the probability of failure increases exponentially when the system is in the red zone.

DRAWING FLOW LINES

One useful technique for modelling a site is a crowd flow diagram. Take a site map, use a sheet of baking sheet/tracing paper and drawing lines on your design to indicate where the crowds are coming from and going to. Draw all the lines where you think that crowds will form and move.

Three characteristics are indicative of risk in this diagram. When lines come together to a point (convergence), where the lines are crossing, or where the lines are in opposite directions (counter flow). By drawing the crowd flow lines on the paper, if you see convergence, crossflow, or counter flow, the risk of a trip, slip of fall, will be high in a high-density crowd. You can use any graphics overlay on any map, plan, or diagram. The result is a risk map for crowd flow. You can use thicker lines for high expected crowd flow, thinner lines for a lower crowd flow, the red area where there are

[37] The instrument that measures the number of revolutions per minute on a car engine.

crossing points, orange/amber for moderate density/flow (50%–80% of the safe density or flow) and green to indicate this area is less than 50% of the safety limit (density or flow). The result is a picture of your potential risks relative to crowd density and crowd flow.

DENSITY AND FLOW IN THE GUIDANCE

The guide to Safety at Sports Grounds (Green Guide) uses the value of 66 people per meter of a minute on stair or slopes surfaces greater than 1 in 20 and on level surfaces, 82 people per meter per minute. Section 10.9 of the 6th Edition for the Green guide states the following:

> The flow rate, also known as the rate of passage, is the number of people who can pass through a particular point in a given time and forms a fundamental part of the calculation of the capacity of the exit system

Flow rates are critical to understanding the impact of crowd risks in high density. Section 10.10 of the 6th Edition for the Green guide states the following:

> For the purposes of calculation, flow rates are expressed in terms of persons per metre width per minute. For an exit route the maximum flow rates used for calculation should be:

a. On a stepped surface, 66 persons per metre width per minute
   Thus, with an exit route measuring 1.2 metre in width, 79 persons should be able to exit reasonably in one minute.
b. On a level surface, 82 persons per metre width per minute
   Thus, with an exit route measuring 1.2 metre in width, 98 persons should be able to exit reasonably in 1 minute.

The values are from the BS EN 13200-1:2019[38] (and previous European standards) which outline units of exit width. 98 persons per minute across a threshold of 1.2 metre=81.67 people per metre per minute (98 divided by 1.2=81.67). There is a caveat to the SGSA guidance in that:

> It is emphasised, however, that basing the design of a level circulation route on a flow rate calculation of fewer than 82 persons per metre width per minute will achieve more comfortable circulation.
>
> Note also that flow rates may be affected by changes of direction along the exit routes, and, in the case of stairways, may be reduced if the movement is upwards.

[38] https://shop.bsigroup.com/ProductDetail/?pid=000000000030240520

With a warning that raising the flow rate is not acceptable

> It is stressed that raising the flow rates used for calculation in order to meet exit or emergency exit times is not acceptable.

ASSESSING FLOW RATES

How do we assess our flow rates? We first determine the narrowest point in the system. The pinch-points or bottlenecks, where the density might be increasing and where your crowd flow might be slowest. That is the point that will determine the overall system flow rate. We would also assess the route complexity, the number of turns, the points that people may stop and wait, where there is an opportunity to take a photograph. Those bottlenecks will determine the overall flow rate through your system. You should calculate your flow rate at that upper limit, 82 people per meter of a minute. That is the maximum, the best, you can achieve. Then, you scale that value down for complexity, congestion, and human factors.

WHAT FACTORS AFFECT FLOW RATE?

Firstly, there will be a different flow rate for ingress and egress, the crowd might use the same design for ingress and egress, but they will arrive over a longer time than egress (normal and emergency). When the crowd leaves a site/event/venue, they might all be leaving at once. Factors such as density, different for ingress and egress, areas, weather, terrain, slopes, stairs, barriers, noise, information systems, crowd management there are several factors. We always measure sites, take sample counts, to give us an idea of the values for specific sites. If you are not sure, measure your site, record your flow rates, then revisit your risk assessment.

SUMMARY – UK AND USA

The crowd dynamics are the same, the world over. The anthropomorphic sizes and demographics of the world's population, the increase in body sizes over the last decade, the pandemic and social distancing, all influence safe crowd density calculations. You calculate a safe space based on all these factors.

You can use the 1, 2, 3, 4 and 5 people per square yard/metre and a visual indicator of density and flow, for control room evaluations of density and flow. The objective is to prevent the crowd density exceeding the safe limits and avoid overcrowding and to avoid the significant risk to life and limb.

Once your space exceeds the safe density limits, whether a small queue is waiting outside a nightclub or a mass gathering, the localised crowd density can cause injury. High density is the crowd risk you need to avoid.

SHOCK WAVES

What happens to the crowd dynamic at high density? In the classroom exercise, we used the square metre marked on the ground, put 1, 2, 3, 4, etc., people into the area, and, with permissions, we would push on one side of the group. Nothing much happens until the crowd are in close contact, then we see an amplification of the push. This is a demonstration of a shockwave. The force is multiplied by the mass of people adding to the momentum of the initial push.

The explanation involves a lot of physics, but a simpler demonstration is the "domino effect" where a domino, standing on its end, can knock over a domino 1 ½ its size when toppled. There are several videos and animations of this on YouTube. One of the best[39] shows slabs of concrete, in a gym, being knocked over, much to the delight of the demonstrator. Another shows the world record[40]. The physics of the domino collapse indicates the forces that are generated when objects topple onto each other.

SIMULATIONS

In our courses, we show a clip of a crowd entering a station – first, with the sound off. Then we ask the delegates what did they see? Most answer with "just a crowd moving into the gap". Then, with the sound on, a remarkably separate set of answers. The sound of shouts and screams and obvious distress coupled with the image gives a clear indication that something is very wrong. Control room CCTV is silent, and we can miss early warning indicators of a crowd in distress, delaying the initial response times. The video clip showed a shockwave, and how it starts, how it forms and how it moves through the crowd. We can see the pressure increasing in seconds.

CROWD PRESSURE

In previous paragraphs, we outlined the UK and USA references to density and flow. Crushing is not the primary cause of death in crowd related incidents. From Fruin's paper[41], *Cause and Preventions of Crowd Disasters*

---

39 https://www.youtube.com/watch?v=YXn08b5Illo
40 https://www.youtube.com/watch?v=JcJv712W_h8
41 https://www.gkstill.com/Support/crowd-flow/fruin/Fruin2.html

> Compressive asphyxia has occurred from people being stacked up vertically, one on top of the other, or horizontal pushing and leaning forces. In the Ibrox Park soccer stadium incident, police reported that the pile of bodies was 3 m (10 feet) high. At this height, people on the bottom would experience chest pressures of 3600-4000 N (800-900 lbs.), assuming half the weight of those above was concentrated in the upper body area.
>
> Horizontal forces sufficient to cause compressive asphyxia would be more dynamic as people push off against each other to obtain breathing space. In the Cincinnati rock concert incident, a line of bodies was found approximately 9 m (30 ft) from a wall near the entrance. This indicates that crowd pressures probably came from both directions as rear ranks pressed forward and front ranks pushed off the wall.

A 20% compression of the chest cavity[42] can break ribs – this is around 3,000 N (three thousand Newtons, from medical bone fracture analysis), so we can see that there is a potential for a five-person deep push to break ribs[43]. Using a crowd flow diagram, you can anticipate where pressure might build up, where the shockwaves might start. Armed with that information, you can design safer spaces for crowds, specifically in queueing systems.

Figure 3.15 shows the station queue crowd flow diagram of the station queue.

If your drawing shows convergence, crossflow or counter flow in a high-density crowd, the risk of a trip, slip or fall, resulting in significant risk to life and limb, will be high. Of course, identifying the problem is only the first step. You now need to understand the elements of designing a safer system for this specific location.

DESIGNING QUEUEING SYSTEMS

The problem with the queueing system shown in Figure 3.15 is that the direction of crowd flow, the crowd convergence, is part of the design. It drives people towards a bottleneck. Density increases and there is a risk that if anyone slips, trips, or falls, there will be a crowd collapse. This kind of system requires design, information, and management elements. In the video, the control is the gating (at the bottom of the screen) which opens when a train leaves the platform and closes when the platform is full. This crowd is competitive. The faster they gain access, the quicker they can get home. This design induces a competitive behaviour in the crowd.

---

42 Biomechanics of Chest and Abdomen Impact (David C. Viano, Albert I King) 2000 CRC Press LLC.

43 Kroll, Still, Neuman, Graham and Griffin "Acute forces required for fatal compression asphyxia: A biomechanical model and historical comparisons" Medicine, Science and the Law. April 2017.

Figure 3.15 Crowd queuing outside a station.

SAFER DESIGN

The moment those gates open, there will be a surge. That is the point at which, in high density, the shockwave can develop. This system creates the problem, creates the "free for all", competitive behaviour. We can define this as an induced competitive crowd. The task is to design all the elements required for a safe queueing design. You need to think about what would happen in an emergency. How do I clear that space? You might need to consider mobility impaired. You should consider the information within the system. Design, information and management are your tools to shape the crowd's behaviour to make this system safer.

Consider what happens when the next train arrives? How is that information transmitted to the crowd? Does this crowd have any scheduling information? There does not appear to be an order in the queueing system. Specific areas in this system are moving faster than others. Does this require an explicit form of crowd management? Can we change the design to prevent the pressure from building up?

The pressure is in the same direction as the movement, the same direction that the crowds are converging on, the entrance. There is nothing to prevent that pressure from reaching dangerously elevated levels (greater than 6–7 people per square metre).

DESIGN TASK

From the UK Guide to Safety at Sports Grounds (6th Edition), four people per square metre is the maximum queueing density. We anticipate that 400 people might be queueing at any one time. What space do you need? How should you design your barrier configuration? This is the terminus, so all trains are heading out in the same direction. Incoming trains have a separate exit point and are empty when they arrive. There is a train due to leave every 10 minutes. With this information, you have all you need for a queueing system design. For example, the area required for the system, the number of barriers, hence the cost of the system.

Sit down with a pen and paper and work out what an ideal system would look like. Consider the elements of the design, the information and the management requirement. How do you make this a safer system? Close the book, try it before reading on.

ELEMENTS OF QUEUEING DESIGN

Hopefully, you closed the book and tried to figure this out for yourselves. Figure 3.16 shows a system with consideration for design, information and management of this crowd.

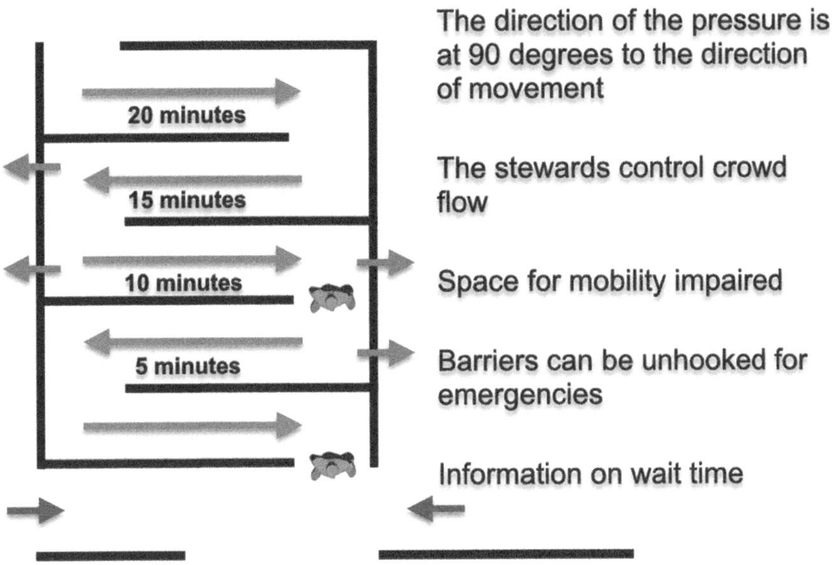

Figure 3.16 Elements of a queueing system.

We have heard this called a *zigzag* or *snaked* queueing system. We can leave a space between the barriers and the station for mobility impaired or VIP ingress and emergency services access. The space is also useful as a contingency space to move the crowd away in an emergency.

This is a "hold and release: system" controlled by the two stewards/ushers shown in Figure 3.16. The upper stewards/usher holds crowd back and the crowd between the stewards (calculated as a trainload using area and density) and releasing them as the train departs the station. The platform, or train capacity, is the number of people between the stewards/ushers.

As you can see, the crowd pressure is at 90 degrees in the final direction (station entrance). The side-to-side movement is important, do not design systems that the crowd moves forwards (top to bottom in Figure 3.16) then back (bottom to top). Psychologically people feel like they are getting closer to their goal, then they are turning and heading away from their goal. The perception of progress is misleading. You want to make these kinds of systems that do not adversely influence the crowd's expectations and behaviour. The design in Figure 3.16 uses a side-to-side movement. This will feel like forward progress, so psychologically, it feels like you are constantly moving towards your goal.

The steward/usher in the middle of the diagram holds the crowds back, while the steward/user at the bottom of the diagram moves out of the way to allow the crowd between the stewards/ushers into the station. The steward/usher at the bottom moves back into position and the steward/usher at the top moves out of the way, allowing the space between the two stewards/ushers to refill that part of the system.

You can uncouple one (or more) barriers from the sides for an emergency access or egress. If you need to clear this space, the crowd does not have to turn and go back on itself.

You can include information within this system. Signage that can tell the crowd how long it is from each section. This is like the Disney[44] style information systems. They tell people how long it is going to take and understand the sunk cost[45] phenomenon. Organisations like Disney measure and monitor the queueing times. They then use signage. For example, from here, it is a 45-minute wait. You can then decide whether you want to spend that time in the queue, invest (hence sunk cost) your time, or reward the ride or attraction.

However, there is a psychological trick they use to manage your expectations and to enhance your experience. As they constantly monitor the crowd flow, they know it is not 45 minutes from that point. It is less, say, 40 minutes.

Here is how that works, you invest your time to get to the front of the queue. You were expecting 45 minutes, it only takes 40 minutes, you feel

[44] https://www.qminder.com/disney-queueing/

[45] https://en.wikipedia.org/wiki/Sunk_cost

good. As they have overestimated the time, or they have measured and calculated the actual time, and added extra, simply to provide an illusion of efficiency.

Let us consider a situation where you post the sign as "45 minutes from this point" and it takes 45 minutes: nothing, no feeling of efficiency. Let us now consider a situation where the sign states 45 minutes and it takes 46 minutes to get to the front. If you are watching the clock, this has an adverse sensation. It is inefficient, poor management, you lied to me, multiple reactions for that extra 1 minute. Now compare that to the illusion of the 5-minute gain. Which behaviour do you want to induce in your crowd? Do you want to manage their expectations? Or do you want them to think you have issues organizing a drinking session in a brewery or other similar analogies?

INFORMATION SHAPES BEHAVIOUR

In our classroom exercises, we will tell the class, if we keep coffee breaks short, and are all back to the class on time, we can all leave early. It is the same effect, influencing behaviour through information, shaping and managing their expectations. We have run courses around the world, the sunk cost phenomena work well, regardless of country, culture or crowd.

DESIGN ISSUES

We often conduct "mystery shopper" site surveys, taking the transport system, as a patron, to events. The aim is to report to the safety manager on anything we can identify, can improve, or advise against. A recent survey for a national stadium identified an area for improvement, in their concession queueing system. Figure 3.17 shows the design.

The crowd had to walk through the entire system to get to the bar. This was one long queue. However, it starts empty, so people had to zigzag up the entire distance. A better design would have been to have the gap so that the first person walks straight through, and as the queue builds up, you close the barriers. Figure 3.18 show the progression of queue build up and barriers closing.

You see this type of queue management at the airport check-in desks. They introduce an extra length of the queuing system so that the first people arriving at the system get through very quickly. Then as the queue builds up, the queueing system expands. You will create greater customer satisfaction and you can get those first customers to the front of the queue much more quickly. This kind of queue management feels much more controlled, more ordered.

Figure 3.17 Concession queueing.

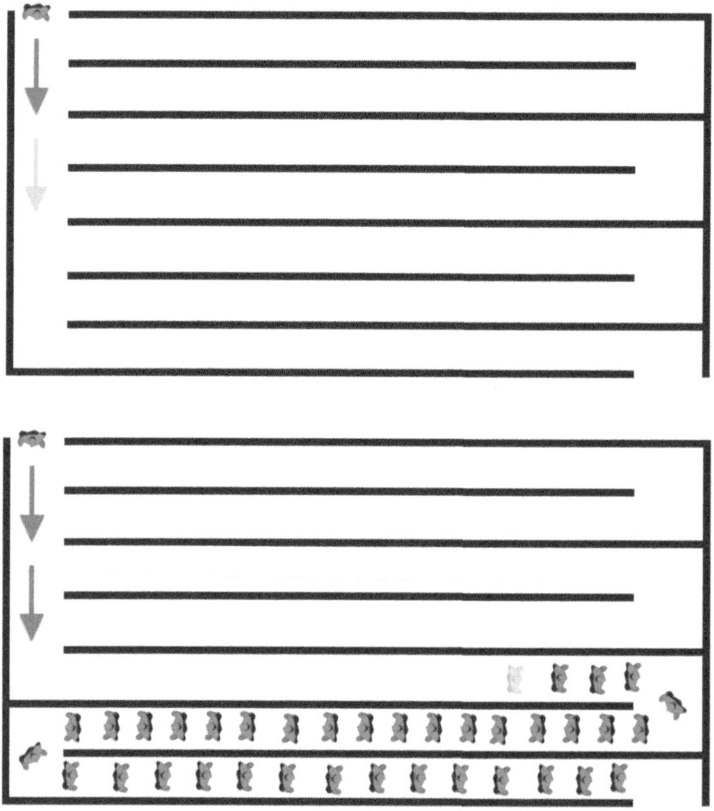

Figure 3.18 Faster service.

YOUR QUEUEING SYSTEMS

You should consider how you design your queuing systems. They shape your crowd's behaviour at the start of your event. What kind of information do you provide? If you are a point of sales, such as a bar or a concession, you want to serve people as quickly as possible. You do not want them walking for 3 or 4 minutes when you could be serving them. Queueing systems design is a particularly important aspect of crowd safety. It keeps the crowd in a manageable controlled environment. If designed correctly, they can prevent the crowd pressure from building up, provide a sense of order, and it helps with the psychology and behaviour during the event. Consider a crowd arriving at a long and chaotic queuing system. It sets their expectations for the rest of your event.

HOLD AND RELEASE

The queueing crowd is stationary, so the density of four people per square meter applies for safe queueing. Hold and Release systems can work for a wide range of crowd sizes. Wembley Stadium, for example, operates a hold and release system for the egressing crowds. Figure 3.19 (circa 1995) shows the police horse operation of hold and release for the departing crowds.

Between each block of people, there is a line of police horses. This is regulating the crowd flow to the station entrance (bottom of the image). The crowd is a station load, the same method as we described above. When the station empties, the next batch of patrons can enter the station.

Figure 3.19 Hold and Release crowd management.

BALANCING THE FLOW

When a moving crowd exceeds a safe density and a trip, slip, or fall occurs, the consequences can be mass fatalities. The "hold and release", using police horses, is regulating the crowd flow on the approach to the station to balance the flow rate[46] out of the stadium to the ingress for the station.

There are systems, specifically with large crowd sizes, that you cannot operate a "hold and release" type system without the risk of crowd pressure building up due to the operations. It is this pressure that can lead to a progressive crowd collapse.

SIMULATING A PROGRESSIVE CROWD COLLAPSE

The author developed a range of crowd simulations, initially focussing on crowd dynamics, then on modelling shockwaves in crowded spaces, knocking simulated crowds over to understand the dynamics of a shockwave through a crowd. This was "crash test dummies"[47] with crowds, which we later developed into a persistent standing crowd collapse model to evaluate the safety of rail seating at football grounds.

Crowd queuing systems can be "hold and release" type mechanism, as with Wembley Stadium. As we have observed, with sufficient space and appropriate crowd management, you can control tens of thousands of patrons leaving a stadium.

However, you need to understand the crowd's behaviour and manage their expectations. For example, the Wembley crowd knows they will have to wait, so management is in line with their expectations.

When there is limited space, or time, or where the density might exceed a safe capacity, you may need to design a free-flowing system, one in which the crowd does not stop. The Jamarat Bridge is one example, and we outlined that project in *Introduction to Crowd Science* and our online courses. We outline the basic principles of flow balancing, designing a free-flowing system in Chapter 8 – Queueing systems.

46 We will explore the calculations for this type of system in Chapter 8 – Queueing Systems.
47 https://en.wikipedia.org/wiki/Crash_test_dummy

Chapter 4

Crowd modelling

> The only thing to do with good advice is to pass it on. It is never of any use to oneself.
>
> **Oscar Wilde**

In this section, we touch on the topic of crowd and event space modelling, using an example from the Dubai New Year event. In 2015/2016, we ran courses in Dubai, for EMAAR, on the topics of crowd dynamics in and around the Dubai Mall. We also focused on the New Year Events. One of the problems they wanted us to evaluate was the crowd management at the station, which is located inside the Dubai Mall. Following the New Year Fireworks, hundreds of thousands of patrons leave at once. This vast crowd was overwhelming the station access points and it was a major discussion topic during the courses. The delegates worked on the crowd management issues around the Promenade, the various floors of the Dubai Mall, the Fountain Show and the station management.

Like other New Year events, the issue is that the crowd will typically arrive over an extended period. The crowds will accumulate around the site, depending on vantage points and transport capacity, weather, and scale of the fireworks display. Then they will all leave the moment the fireworks finish.

For city sites, the transport system regulates the rate of ingress of crowds to the display areas. The station platforms and vomitories, escalators, lifts and stairs, all have sufficient capacity to manage trainloads of spectators arriving over time.

The problem is that the number of people trying to access the station grossly exceeds the system's capacity during event egress.

There is a bus terminal, access through the Dubai Mall and a Metro link station inside the Dubai Mall. Both were subjects we studied with the EMAAR staff. The aim was not to provide an external solution to any problems, but (as with all our courses) to train the staff how to solve these kinds of crowd dynamics for themselves. Their use of graphics facilitated planning communications, as the group has a good command of English, much more than our knowledge of Arabic.

DOI: 10.1201/9781351053068-4

RETRACING THEIR STEPS

If the spectators arrive at a site/venue/event from a specific direction, using a specific route, this will pre-conditioned them to leave by that same route. In general, people go out the same way they come into a site/venue/event.

For this event, the New Year fireworks, the crowds will arrive throughout the day/evening and enjoy the facilities at the Dubai Mall. Then, after the fireworks end, they will all try and leave at once. As there are both train and bus terminals within the Dubai Mall, the passenger egress demand exceeds the transport capacity. Crowds along the routes from the viewing areas to the transport systems become overcrowded. In previous years, they had tried to close the transport systems on egress. This was the main topic for evaluation during our crowd science workshops for EMAAR. As with all our courses, our aim was teaching the delegates how to solve their own site crowd management issues. As each site can have unique features, and each delegate may have unique experience of their site, our job is to augment that experience with the crowd risk analysis tools.

For example, we outlined the issues of closing the transport system for egress only, a strategy that needs to be balanced with alternative ingress/egress options. We reviewed the principles of design limitations, how their information systems (how to get to the fireworks and how to leave the area in normal and emergency stations) and how to manage the crowd during egress. When the Address Tower[1] fire threatened the site, which occurred a few weeks after our courses ended, our planning tools were assessed. There were no lives lost in the incident although there were tens of thousands of people on the site at that time, a successful emergency site evacuation. From: https://www.mirror.co.uk/all-about/dubai-hotel-fire.

> A huge fire broke out in a hotel in Dubai ahead of the 2016 New Year's Eve celebrations. Pictures emerged on Twitter showing The Address Downtown Dubai hotel engulfed in flames. The blaze began hours before a large fireworks display was set to start from the nearby Burj Khalifa tower, the world's largest building.

LONDON NEW YEAR

The Dubai New Year and London New Year events have a similar problem, in that there are mass transit systems close to the event footprint, where the bulk of spectators arrive and depart the event. In *Introduction to Crowd Science*, we outlined the key issues of the site, so present only a summary of this event below. The focus of the fireworks is the Millennium Eye (solid ellipse).

[1] https://www.bbc.co.uk/news/av/world-middle-east-35208484

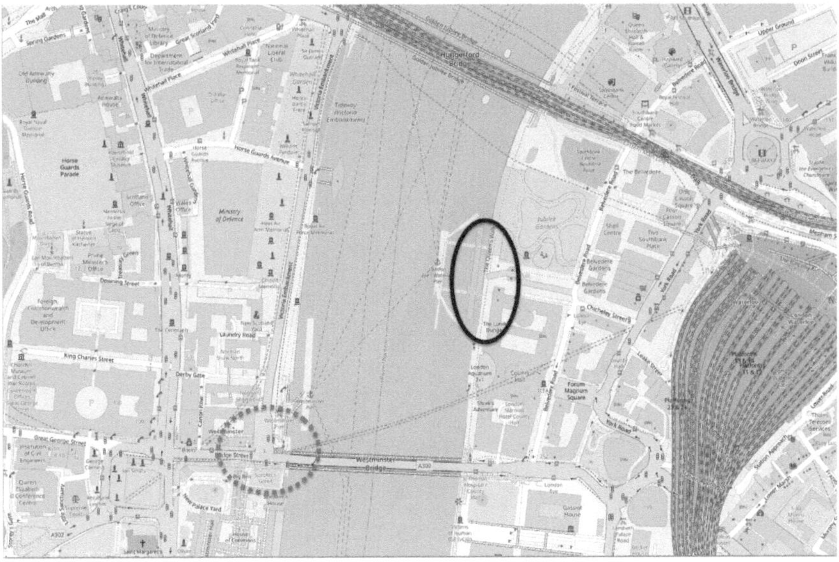

Figure 4.1 London New Year Event.

The area of concern is the junction close to Westminster station (dotted ellipse) in Figure 4.1.

As spectators arrive via Westminster station, they will depart using the same station. The image below shows the congestion (inverted infrared, black is high-density crowds). In 2005/2006, there were problems of crushing at the station entrance and the site was the topic of a workshop at the UK Cabinet Office Emergency planning college.

From the video footage, it is difficult to make out individuals and determine the nature of the crowd issues, such as the crowd direction and speed of crowd movement. This is where your crowd flow diagram will help identify the source of the crowd congestion problem.

We can see the extent of the crowd congestion (Figure 4.2) and the line diagram (Figure 4.3) shows the directions of crowd movement, proximity of the station to the event footprint. This kind of line diagram provides an insight into the nature of the problem. We can see the congestion, in the convergence, crossflow and counter-flow from the line diagram (Figure 4.3). These lines are a mathematical representation of crowd flow and can provide valuable insights during the planning process. Where you see convergence (lines coming together), cross flow (lines crossing each other) and counter flow (lines moving in opposite directions), there will be congestion and overcrowding in high density. Models provide one of two key outputs. They can either prove or disprove an idea, or they can provide insight to the situation that might evolve. The number of convergent, crossing and counter flow, the

Figure 4.2 Helicopter-inverted infrared image.

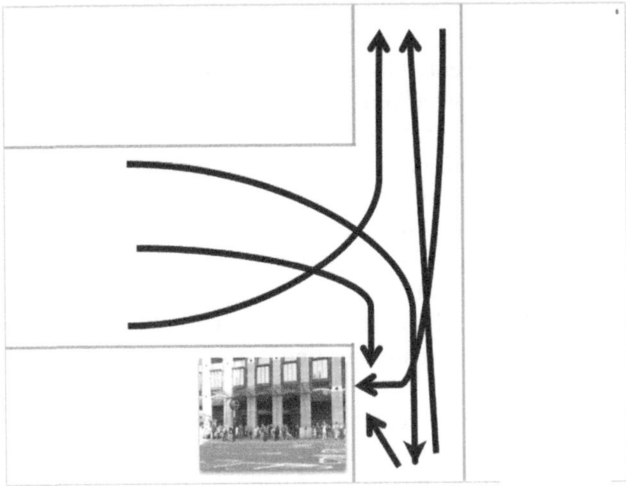

Figure 4.3 Crowd-flow diagram.

number of people moving in this space, the station's location and the station's capacity are all contributing to the risks of overcrowding in this area.

At this site, the station had become overcrowded. To avoid overcrowding on the underground (the Tube) platforms, British Transport Police shut the station gates.

As the crowds moved towards the closed station entrance, severe injuries were due to the high pressure on the system. This incident was the primary motivation to take this case study to our workshops.

In subsequent courses and workshops, we provide the above outline of the problem, then challenge the class, using Design, Information, and Management to make that process (egress from the event, ingress to the station) safe.

The solution is based on shaping the crowds' expectations. People will leave a site (generally) along the same route they entered the site. You can therefore shape the direction and behaviour of the egressing crowd by changing their mode of transport and direction of crowd ingress *to* the site. In other words, this kind of egress problem is actually an ingress problem. The risks are on egress, but the solution is to change the crowd's direction of approach to the site. We are shaping the crowds' expectations for egress, using design, information and management during ingress.

MODELLING THE SITE

We use "baking" sheet (tracing paper) in classroom exercises as the physical act of drawing a line on a piece of paper fires multiple senses. The length of the line (it takes longer to draw a longer line), the colour of the line, the width/thickness of the lines, the number of lines, the visual aspect and the sense of touch. All of this helps the user build a connection with the crowd dynamics of the site/event. Given the environmental considerations, and enhanced graphics of PowerPoint (or Keynote), the result is easier to transmit using computer graphics, but computer's nature (email, social media, and other forms of communication) can be a distraction. When modelling for crowd safety, you do not want those distractions.

When we teach these principles at an MSc level (and via our online courses), we advise using a Google Earth image or site map and import that to a blank PowerPoint (or Keynote) slide; then using the saturation slider to reduce the background to a greyscale. This highlights the colours you add, and then add semi-transparent lines, circles, in the Red/Orange/Green colour scheme.

We always start crowd projects with baking sheets (tracing paper) and the computer switched off, specifically if it is a side project for a distant location. Also, we always recommend walking the site (if possible) as you will notice three-dimensional detail you may not see on a plan. Finally, make sure you check the site as close to the event as possible. We have seen last-minute roadworks change a route width or block it off entirely.

EGRESS IS A FUNCTION OF INGRESS

If we can dictate the route the crowd will use to approach a site; we are preconditioning the crowd to go back out via that same route. The problem we see on this site (and on other similar event sites) is that the station is close to the event footprint. So obviously, it will become an attractor; people will want to go to the site via the shortest route. However, if we shut the station prior to the event and force them to come in another route, they will return along the route they used for entry. We also need to inform the public that the station will close for the event and will not reopen. This is shaping the crowds' expectations and hence their egress behaviour. We are conditioning the crowd to take a much safer option.

The other element of the final design for this site was to close off the bridge, at the area closest to the station using a barrier that can easily open in the event of an emergency.

In this manner, we have split the site into two areas. One serviced by the top (Waterloo Station) and the other from stations to the bottom of the diagram above (Victoria Station). These two crowds are marshalled to fill their respective spaces as evenly as possible, making the maximum use of the viewing areas. When the event is over, both areas/crowds exit without any convergence, crossflow, or counter-flows. From an operational perspective, we could open the station when we feel it is safe to do so.

However, this is important in shaping the crowds' expectations. We do not inform the crowd that this may happen. If we tell the crowd the station might reopen, they might decide to wait close to the station entrance, and the area may not clear. As we outlined in the DIM-ICE chapter, information shapes the crowds' behaviour and information changes the crowds' expectations.

As we have shown in these two examples, assessing the ingress/egress behaviour and site modelling (the line diagrams) helps us analyse the situation, define the problem, and that process helps to reveal the solution.

Chapter 5

RAMP analysis

> Knowing is not enough; we must apply. Willing is not enough; we must do.
>
> Johann Wolfgang von Goethe

The DIM-ICE risk model defines the event character and risks based on the analysis of previous major incidents and the underlying primary causes of those incidents. The RAMP analysis defines the crowd dynamics as they approach, move around and depart the site.

FIRST-PASS APPROXIMATIONS

Before the invention of the pocket calculator, or an app on smart devices, we used a slide rule (Figure 5.1) for complex calculations.

The slide rule[1] did not require batteries or an internet connection. It was reasonably accurate to three decimal places. However, you did need to know the order of magnitude of the answer. You had to have a clear notion of the number of decimal places, or zero's you need to add to your answer.

Although there were computers used during the Apollo Moon-shot, the bulk of the calculations used slide rules. We joke that NASA Apollo safety systems, to send men to the moon and back, was based on rubbing two bits of plastic together.

One technique used by NASA, called course correction, has a significance for crowd safety. Course correction knows how much you need to nudge a thing to get it back into alignment when a system starts to drift away from the desired path. You may experience this every time you try a hotel shower for the first time. If it is too hot, you turn down the "hot" and turn up the "cold" (or rotate the temperature knob). If it is too cold, you turn up the "hot" again until the water is the right temperature for

[1] https://www.math.utah.edu/~alfeld/sliderules/

DOI: 10.1201/9781351053068-5

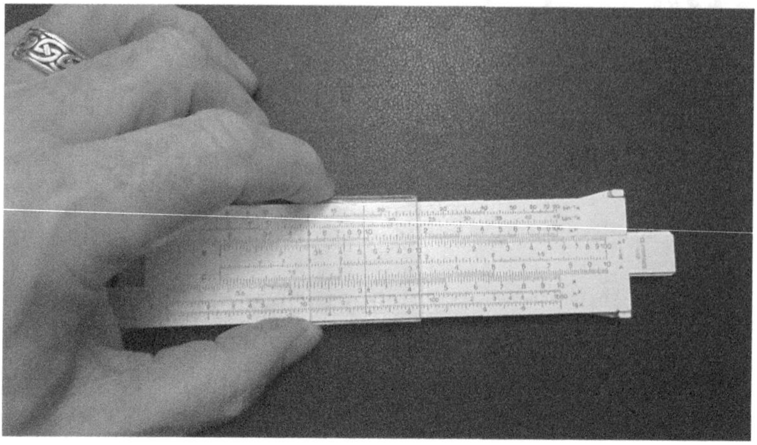

Figure 5.1 A slide rule.

you. You will have noticed that it can take a few seconds for the temperature to change because of the amount of water between the dial and the showerhead, so the reaction of the temperature lags the action of turning the temperature up or down. It takes time for the water to reach a new equilibrium, to settle down.

From our previous chapter, if you notice the density is increasing to the 80% level, in any specific area, you can manage the crowd to prevent further increases, to prevent overcrowding, before that area is at risk. Of course, you need to understand density and flow to identify those rises in density or flow. We can process visual information faster than interpreting written information. Studies[2] show that 90% of the information processed by the brain is visual and that it takes only 13 ms for the human brain to process an image and that the human brain processes images six times to six hundred times faster[3] than text. Furthermore, 80% of people remember what they see, compared to 10% what they hear and 20% of what they read.

Using a Green, Orange/Amber, Red classification can help understanding crowd risk analysis. Using graphic representations of crowd dynamics in the control room helps with the early identification of crowd problems.

Green=0%–50% of the system's capacity, or flow rate. Orange/Amber=50%–80% of the system's capacity or flow rate and at 80%, you close the system, allow the system to settle down and then you can evaluate the situation.

[2] https://www.t-sciences.com/news/humans-process-visual-data-better
[3] https://www.emailaudience.com/research-picture-worth-1000-words-marketing/

WHAT IS THE BEST WAY TO PLAN FOR A NON-TICKETED EVENT?

Over the last few decades of teaching crowd science, we have observed, reviewed, and advised on hundreds of events[4]. One of the most frequent questions asked is:

> What is the best way to plan for a major event if we do not know how many people are coming?

For non-ticketed events, like the 150th Anniversary, Canada Day celebrations in Ottawa, the number of people who want to attend may be unknown. So how do you plan for uncertain numbers on your site?

We can work backward from the site capacity, then determine the upper limits, like the slide rule: if we know the rough order of magnitude (numbers of people that the site can safely accommodate), then we can devise a crowd management plan to keep it within those limits.

FOUR CONSIDERATIONS

We only need four parameters for the rough-cut capacity analysis for our initial first pass at our crowd management plan. The first is direction. What direction are the crowds approaching or departing the area? What ***routes*** will they take to arrive and leave the site? This breaks down into the modal split or modal share[5]; which modes of transport will be used? For Canada Day, in Ottawa, it would be the OC Transpo[6] (rail and busses) and cars. So, by knowing the station and bus drop-off/pick-up points, the site entrances and exits, the car parks, and the relevant capacities of each, we know the routes and can work out maximum capacities on each. That will provide an upper limit for the number of people that may travel to the site and the routes they will take to get to and from the site.

The second parameter is the site ***area***. How much space will be available for the crowd? Will your areas be mostly for static crowds, or will they be dynamic in nature, crowds moving around your site? From the previous chapter, you should have an idea of how to divide your site into the statics, dynamic, and contingency spaces.

Routes and Areas can give provide the upper limits of people that can arrive at your site, and if that number may exceed your site capacity. You then know if you will need crowd management to prevent overcrowding.

The third parameter is how quickly those spaces will fill over time? What kind of crowd ***movements*** do you expect? Is it going to be an early arrival

[4] https://crowdrisks.com/expert.html
[5] https://en.wikipedia.org/wiki/Modal_share
[6] https://www.octranspo.com/en/

or a late arrival? Events like football matches or fireworks events, where people are already familiar with the time it can take to get to the site, will plan the journey accordingly.

In fireworks, sporting events, fixed seating events may typically see late arrivals, while celebrity appearances and non-ticketed events may see earlier arrivals. If you have an idea of how the crowds are moving to your site over time, you can plan accordingly. Same for departures, will the crowd all leave at once, or will there be a staggered departure? An estimate of the time that the site is filling, or emptying is important for your contingency planning and crowd management.

The fourth parameter is the ***people***, the profile of the crowd. What do we know of this crowd, their previous behaviour at this site/venue/event – or comparable sites/venues/events. Then you can determine the style of crowd management required for your crowd?

RAMP ANALYSIS – EXAMPLES

The routes, the areas, the movement over time and the people/profile of the crowd are the basics of your planning. Determine the upper limits of the site capacity from the areas you have available. Then the capacity of the routes to and from the site. Then the movement over time and the people/profile of the crowd. With those upper limits, you can determine if the site has any problems "lying in wait".

In *Introduction to crowd science*, we outlined the use of the RAMP analysis. To avoid repeating the previous publication, we will summarise the applications here. There are examples of how to use these models in the case studies, and we have more detailed explanations in the online courses.

ROUTE DIAGRAMS

We always encourage learners to use pen and paper to draw routing diagrams. This facilitates learning because you are using more than your mouse finger to create the diagram. When drawing lines on paper, you have the sense of touch, of time, the time it takes to move your pen from one point to another, of sight and intensity, by colouring the lines red, orange/amber, or green, by drawing thicker lines for higher crowd flows, thinner lines for lower crowd flows. This engages more of your neural activities, making the process of plotting crowd dynamics easier to understand. You can always create computer graphics *after* you have created a drawing, and you will see examples in the case studies.

We recommend you switch off the computer and use a pen/pencil to create your routing diagrams. The process is more intuitive, with less distractions from email, social media and funny clips of animals sent by friends who have little else to do with their time.

MCFC VICTORY PARADE

We were engaged in evaluating the Manchester City Victory Parade in 2011[7]. Figure 5.2 shows the route. The plan was to use barriers (top left of Figure 5.2), along the length of the route, to separate the fans from the team coach. As we walked the route, we highlighted that the crowds would approach from the transport system, from the West (the right-hand side in Figure 5.2). The crowds would fill one side, and the far side (the East, left hand side) would be empty. There was a need to break the barriers, to actively manage the crowd flow to fill the East-/Left-hand side first, then the West-/right-hand side.

The proximity of the station, to Old Trafford (bottom left of Figure 5.2), would also require crowd management as the stadium was not in use, only the areas around the stadium. Normally, crowd flow is regulated between

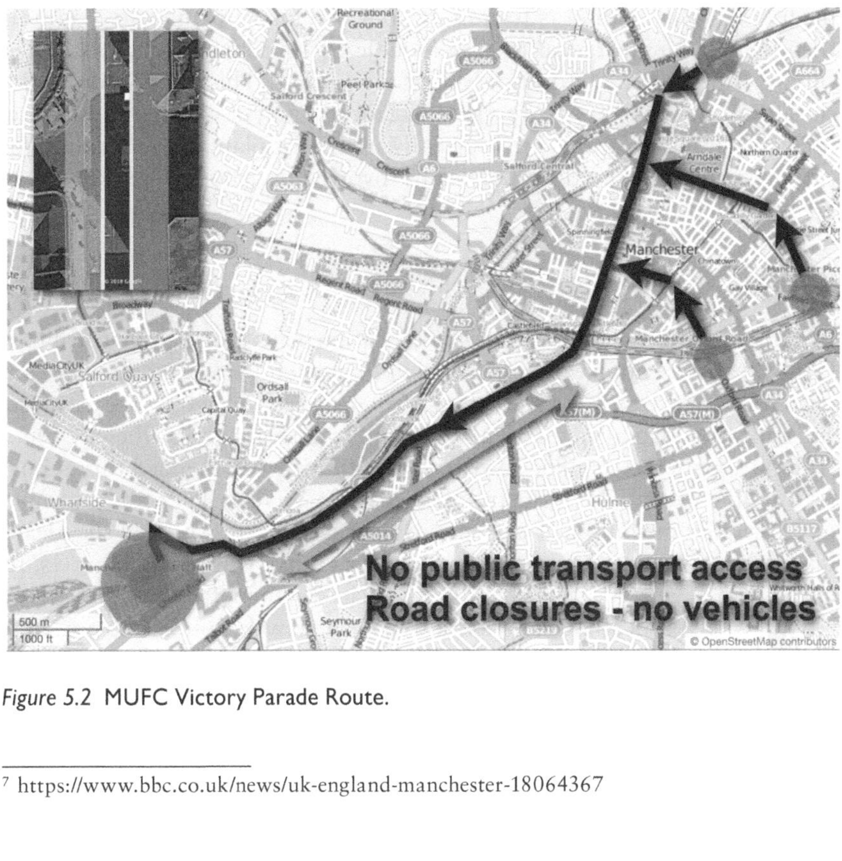

Figure 5.2 MUFC Victory Parade Route.

[7] https://www.bbc.co.uk/news/uk-england-manchester-18064367

the station and stadium, by the train capacity in ingress and by the stadium exit gates on egress. The streets to the station would need additional barriers for the 24,000 fans that might accumulate around the stadium. We advised the Old Trafford station should not be used, it was too close to the event footprint, for this event. We also advised to open the car parks as the stadium staff had experience of crowds and traffic management. We outlined the complete scope of the site issues and solutions in both *Introduction to Crowd Science* and on our courses.

DRAW SIX ROUTING DIAGRAMS

We advise that you create six routing diagrams. A pair of diagrams for ingress (normal and emergency), a pair for circulation (normal and emergency), and a pair for egress normal and emergency. These will provide insights into any conflicting crowd flows if there is an emergency, and you need to change the direction of the crowd flow. These diagrams, when overlayed, can give you insights into the crossflow, counterflow, and convergence on the site. If you are using a computer (with your email and social media off), one method is to import your site maps, plans and diagrams, then reduce the saturation to zero (double click on the image, format picture, picture colour, saturation slider down to 0). This allows you to add coloured lines that are more visible against a grey scale than a coloured image. Routing diagrams are especially useful for planning and communications among the multi-agency teams involved in major events.

AREAS

In our short courses, we cover the site relocation plans for Leicester Caribbean Carnival as the example of areas, both in *Introduction to Crowd Science* and our crowd safety course. The issues of that project related to the proposed site relocation, from Victoria Park (90,000 square metres) to the city centre (10,000 square meters). For 25 years, 100,000 patrons attended this annual, high-profile event. The infrastructure costs of Victoria Park and examples of other cities moving their events into their city centres sparked the idea of relocating this event. Prompted by other city events, Christmas Markets, for example, can generate tens of millions of city revenue, from visitors, across a wide range of local businesses. However, there were concerns over safety. A delegation of first responders (police, fire and first aid) took the site to a workshop we ran at the UK Cabinet Office Emergency Planning college, and we assessed various site issues. There had been a consultant's report produced on the problems, but the city officials were in the mind set of

> Don't come to us with problems, come to us with solutions.

The politicians may not understand the safety consultant's jargon or numbers or may choose to ignore that issue. If we consider that the city centre has 10,000 square metres of space, and that there is an expectation of people coming and going throughout the day, it can look like a crowd management problem.

To increase the spending, they wanted to position concession stands (reducing the available area to 8,400 square metres). They also wanted a parade to run through the site, further reducing the available space for the crowds to 6,000 square metres. Victoria Park had 90,000 square metres for 100,000 people. You can see the potential problems, 100,000 people, passing in and out of the site is difficult to visualise. There was no clear indication of how long they may dwell within the event footprint. As far as the politicians were concerned, this is a crowd management problem. It was up to the team to solve the problem, not harp on the problem's scale, just fix it. We are sure that everyone reading this had had similar experiences. Safety does not make money, requires investment with no return, so it can be a struggle to convince the purse holders to spend money.

Whenever you need to convince a politician or financial controller that you require changes, you need to focus on how *they* interpret information. The numbers, the reports, and the dynamics are not within their areas of expertise. Therefore, they will not equate the consultant's report with a risk to themselves. So, how does a politician, or financial controller, perceive risk? In newspaper headlines, in lost votes, in the cost of failure, the potential lawsuits may follow an incident. You can use a visual *sleight of mind* to communicate the problems to the city officials.

Despite the team working through the issues of the Leicester Caribbean Carnival, the problem was in presenting the consequences of failure in a way that forced a political decision. So, think about "tomorrow's news headlines" about the impact that will have on their constituency and voters' perceptions. That is how to present your concerns.

DO YOU WANT TO MOVE THIS INTO THAT?

A simple graphic, which shows the relative sizes of the two areas, and the threat that "this will be tomorrow's headlines on the local and national media channels" if there are any problems. By showing the two alternatives side by side, to the same scale (Figure 5.3), the visual comparison, the *sleight or mind*, plays to our desired effect.

The visual comparison of these two areas becomes obvious and apparent. Do you want to move this into that? It clearly will not fit! Furthermore, we highlighted that if there is a problem with crowd crushing at this site, *this* is the graphic that will be in tomorrow's news. We showed you the problem, in a way that even a child can understand. It is

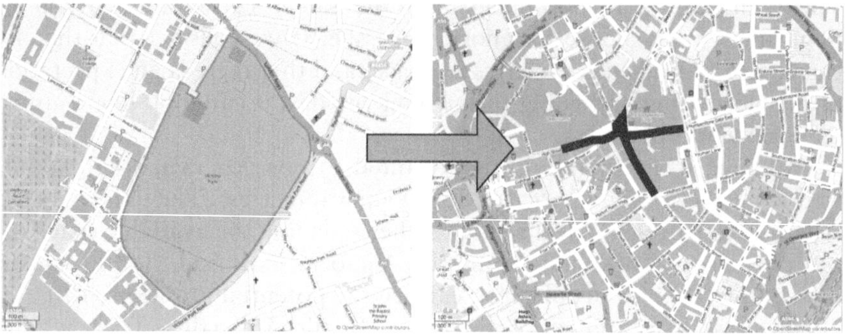

Figure 5.3 Comparing the site areas.

the visual sleight of mind, the psychology, which forces the decision to the safer option. This is straight out of the magician's toolkit, letting the mind fill[8] in the blanks[9].

SITE MEASUREMENT

Measuring site areas and distances, using tools like Google Earth Pro, and our own tool, called the 3D RTDS (Real-Time Decision Support) is the starting point. We can point and click over an imported map/plan or satellite image when using these tools, to measure your spaces and distances.

We can then obtain an area (available space) from these tools, but before you use that to assess site capacity, deduct 20% from the reported area. There are two reasons for this; the first is that the maps may not be accurate, so build in your margin for error *before* you do your event capacity calculations. Then you can deduct space for proposed infrastructure, such as barriers, toilets, front of the stage, back of the stage, concessions. You can always re-measure the site once the plans are finalised and upscale your capacity.

If you start with a small number, this shapes the expectations of the event organiser; and raising capacity is a bonus, more patrons and more revenue. If you start with a high number, and then the need to lower your capacity (because you missed X, Y, Z on the site), it will result in fewer patrons and hence lower profit for a site/venue/event. There may be resistance to reducing capacity. Information shapes expectations across the spectrum of event planning.

[8] https://www.scientificamerican.com/article/filling-in-the/
[9] https://www.sciencedaily.com/releases/2000/06/000601164617.htm

MOVEMENT

You have worked out your areas (and built in your safety margin), you know your routes, their widths, and hence the maximum flow rate that those routes can accommodate (82 people per metre per minute times the width of the route, or 25 people per foot per minute times width of route). You have defined the 50% and 80% limits (for both flow rates and site filling). Your crowd management plan is beginning to take shape by evaluating the geometry of the site, the maximum capacity of the site, the routes, the areas used.

We need to know the times that people will try and get to the site. If this is a green-field (new) event, then the upper limit will be the transport capacity. That will be the maximum arrival rate. Are your crowds going to come early or will they come late? There are three fundamental shapes of the arrival profile, shown in Figure 5.4.

The downward curve is typical of an early arriving crowd. This will taper off as time approaches the start of the event. The late arrival crowd, where the curve moves upwards as the time approaches the start of the event, and the stepped (up and down) function, which is a limited transport arrival, e.g., trainload every 10 minutes.

Knowing your crowd arrival profile, the movement of people over time, will help with your planning and resource deployment for staff and security at entry gates. You can see how you can plan the number of turnstiles,

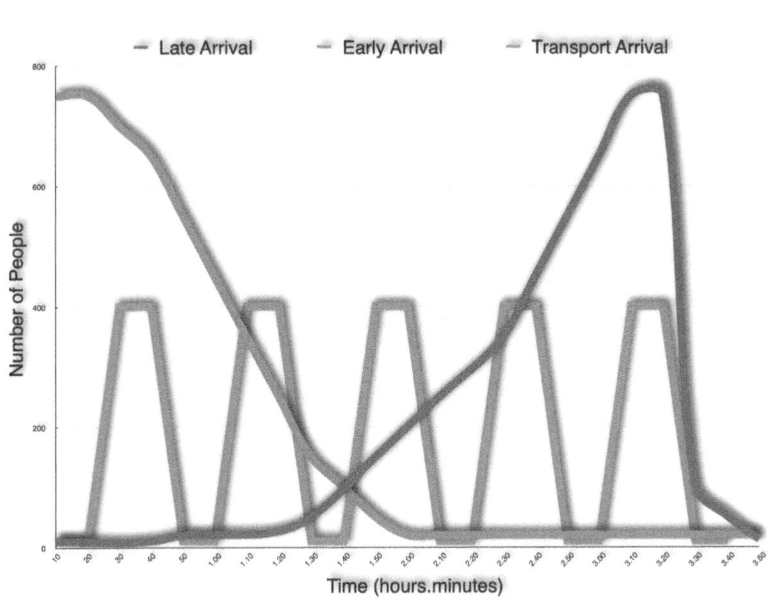

Figure 5.4 Arrivals over time.

security checks, and calculate the time to fill the site from the site geometry and RAMP analysis. If the arrival rate exceeds your ability to process tickets or security checks, then queues[10] will develop. Our MSc students have used the arrival profile to improve resource planning and site design pre-pandemic. During the Covid-19 Pandemic, the evaluation of queueing's, with social distancing, vaccination checks, trial events and changes in the security/screening. Again, MSc students were engaged in systems design, using queueing theory.

ETIHAD STADIUM PROJECT

In 2014, we were consultants for the crowd management during the South Stand development at Manchester City's ground, Etihad Stadium. During the construction, they knew they would have a reduced number of turnstiles for football matches and concerts (Figure 5.5).

From the turnstile data, we knew the typical arrival profile the impact this would have on fewer operating turnstiles. This is a typical sporting event arrival profile. Spectators understand how long it takes to get to the stadium. They understand the transport system capacity, and they have

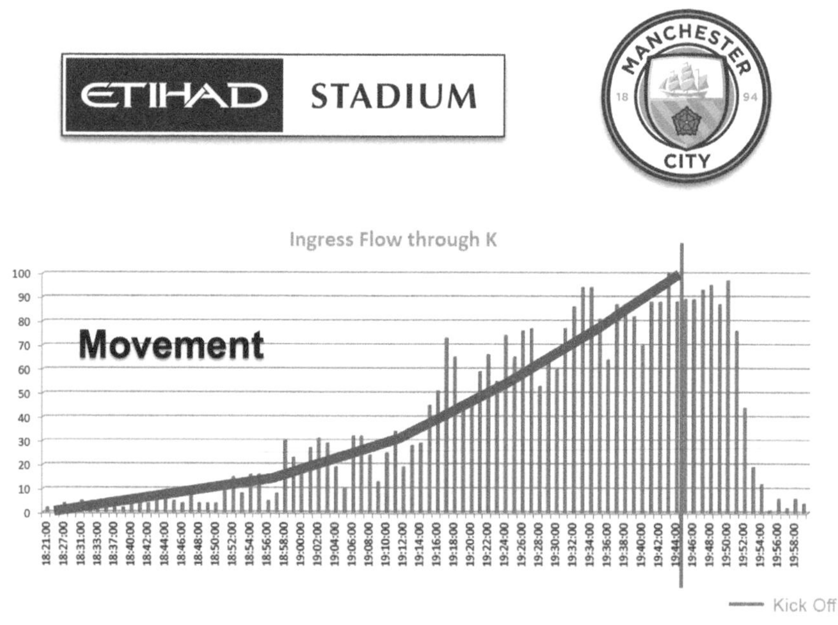

Figure 5.5 MCFC arrival profile.

[10] We cover queueing analysis in Chapter 8.

been coming to the stadium for years. We needed them to come earlier, to avoid congestion at the turnstiles, close to the start of the match.

We can calculate that, if we can shift the arrival profile by five minutes, get spectators to arrive five minutes earlier; then the reduced turnstile throughput would cope with the time shifted demand. The problem is that "five minutes earlier", feels trivial, not important; therefore, behaviour may not change. We needed to shape this crowds' behaviour, to manage their expectations. The information needs to have an impact, to feel that an effort is required.

Through the stadium on the supporters' websites, we presented a series of public information announcements on the transport system and via the local media. First, set the expectations; this is the new site design (images and video of the new South Stand). Then set the travel expectations. During the build, please allow an extra twenty minutes to get through the turnstiles. Twenty minutes feels like you will need to make effort. Not everyone will do this, some will ignore it, some may forget about it. We factored that we only required 5 minutes into the analysis and knew how the averages and human behaviour would react to the scale of 20, compared to the scale of 5-minute delay.

We have used that technique on events before, and after the Etihad Project. Understanding the level of effort that shifts travel expectations, hence the crowd's arrival profile, and reduce the potential for overcrowding, is another example of sleight of mind. We know there will always be a minority who are just going to carry on the way they were or did not see the message, or did not understand what it meant, or forgot on the day. However, we can shift the behaviour of enough spectators to eliminate the problem, to take the pressure off the turnstiles in those last few minutes.

2006 LONDON NEW YEAR

The crowd moved over time for a typical arrival profile for a fireworks event is the exponential curve[11]. You can encode the function into an excel spreadsheet and, with real-time measurements, sampling crowd flow, you can use excel to predict when an event will reach 80% capacity. We created a system, in 2006, for the London New Year Event (LNYE). There was a problem on egress, and we outlined in Chapter 5, which we can resolve by directing the crowd during ingress. At the time of this project, LYNE was a non-ticketed event. With a large catchment area for the population of London (approx. Seven million at the time of this project), the site capacity was 186,000 patrons, Figure 5.6 below.

[11] https://en.wikipedia.org/wiki/Exponential_growth

Areas (see image)	*Area (m2)*
A – Waterloo to Embankment Station	11,955
B – Embankment to Westminster Station	23,304
C – Parliament Square	2,092
D – Westminster Bridge	4,138
E – Belvedere Road	11,241
F – Royal Festival Hall	3,000
G – Waterloo Bridge	9,143

Figure 5.6 LNYE areas.

The project involved reducing the number of entry points, real-time crowd counting, using a clicker count and sampling every 15 minutes, then feeding that information to a central control to add to a spreadsheet, Figure 5.7.

The fill predictor model gave the command-and-control team advanced notification of when the site would reach its 80% capacity. With that information and knowing there would be a crowd already on the transport system, we could send information to all stations advising "site full" before it became overcrowded.

The messaging, as with the Etihad Stadium messaging, needs to manage the crowd expectations. In previous events, the management team had

Flow Estimator

Density per metre square	Area Name	Area	Total Capacity	Exponetial Time to Fail	11:15:59 PM	Arrival Profile		Use Street Counts
4	B	23304	93216	Linear Time To Fail	11:54:47 PM	Default	+ 15 Mins	Street Counts Update

Time Period	2100-2115	2115-2130	2130-2145	2145-2200	2200-2215	2215-2230	2230-2245	2245-2300	2300-2315	2315-2330	2330-2345	2345-0000
Arrival Profile	30%	30%	30%	30%	30%	40%	45%	50%	60%	70%	80%	100%
Desired Flow For Period	4700	4700	4700	4700	4700	6267	7050	7833	9400	10967	12533	15667
Desired Aggregate Flow	4700	9400	14100	18800	23500	29766	36816	44650	54050	65016	77549	93216
Overall Street Count	8000	0	0	0	0	0	0	0	0	0	0	0
Predicted Flow Exponential	8000	8000	8000	8000	8000	10667	12000	13333	16000	18667	21333	26667
Aggregate Flow Exponential	8000	16000	24000	32000	40000	50667	62667	76000	92000	110667	132000	158667
Predicted Flow Linear	8000	8000	8000	8000	8000	8000	8000	8000	8000	8000	8000	8000
Aggregate Flow Linear	8000	16000	24000	32000	40000	48000	56000	64000	72000	80000	88000	96000

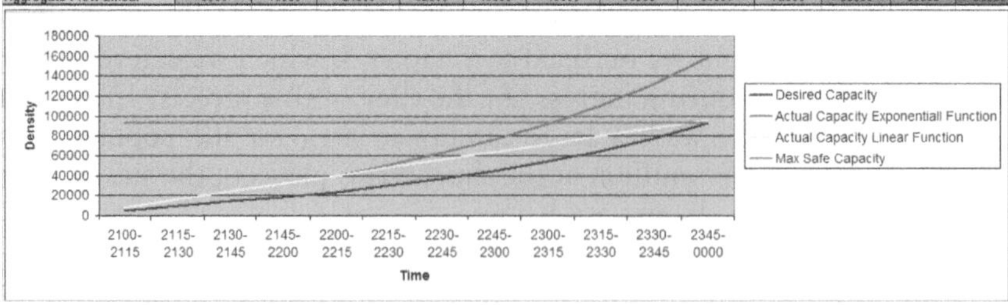

Figure 5.7 LNYE fill predictor.

closed entry points when the site was full. This meant that the crowd on the transport system was expecting to gain entry, then denied. 50,000 people wanted to but could not access the site and there were public order problems outside the event footprint.

To avoid repeating that situation, using the fill predictor, we could manage the crowd expectations, telling them the site was "full" before it reached capacity. In other words, we assessed the number of people on the transport system, how full the site was (from the fill predictor) and by deducting the number in transit and the number on site, we could identify when to transmit the "site full" messages. Of course, the site was not full at that time, but it would be by the time the crowd in transit reached the site. The information must shape the crowds' expectations. If we told the crowd that:

> The site may be full by the time you get there. You might not get in.

How might the crowd react? The announcement is a confusing message and will do little to manage the expectations of the crowd. The announcement that the,

> The site is full

It is clear, concise, and sets the expectations of the crowd. If they do get in, then this is a bonus, but they do not *expect* to get into the site.

We teach the MSc students how to build these models over a five-week class. The system is in use at several international sites and, in 2022, will be used for Canada Day in Ottawa, which is relocating[12] away from Parliament Hill.

If you have an idea of your crowd/site arrival profile, the curves shown in Figure 5.4, you can take a sample crowd count and predict, with reasonable accuracy, when your site will be full. Measure the width of the route you want to sample count, use a clicker/counter and count how many people pass in a minute. Divide by the width of the route and you have a flow rate. Repeat as often as required. Every 15 minutes provides an approximation of your arrive profile, your M in the RaMP analysis.

Of course, ticketing an event ensures that there will be no overcrowding. However, street events are non-ticketed and knowing the arrival rates in real-time provides valuable command and control information. If the arrival rates are different from your expectations, then you can "red flag" the situation. You have identified a problem in a timely manner. You will have time to act, to manage the crowd flow. The fill predictor is a proactive, not a reactive system.

[12] We are currently consulting with OC Tranpo and Ottawa Police for 2021–2022 events.

PEOPLE/PROFILE

What do we know about this crowd, at this site, at other comparable sites? What is your expected demographic, age range, the ability of your crowd? Do you know the history of previous crowd behaviour at this site with this crowd? Those are the "P" questions for the RAMP analysis.

Behavioural psychologists have different classification of crowds[13]. Berlonghi (1995), Pines and Maslach (1993), Zietz (2009) are examples of crowd mood classifications. The student essays discuss these, along with their source's references.

You might create your own classification for your site, your event. What crowd behaviour do you expect? Can you manage the crowds' expectations? It may be that a particular venue attracts a particular type of crowds for particular types of concerts. A recent case in the USA, where we were expert witnesses, had a specific expected crowd profile. The organiser failed to provide a suitably safe environment for that type of expected crowd behaviour. They did not win that case.

SOCIAL MEDIA CROWDS

Be aware that social media celebrities can attract huge crowd – an event in Birmingham[14], UK, Jan 2019, saw a YouTuber, a make-up artist, schedule an appearance at the Bull Ring (shopping mall). Tens of thousands of his followers arrived, some were hysterical, the site was overwhelmed and reports indicate there were significant challenges for managing this crowd.

The YouTuber was James Charles, at that time he had 16.2 million followers. There was pre-event information about a large crowd turning up to the site, and some significant issues associated with crowd management were observed. In our courses, we show a video of this site and the pressure on the mall barriers. Any sudden surge and there could be a significant concern for crowd safety from a barrier collapse. From the shockwave models (Chapter 3), we can assess the risks of barrier collapse. There was no major incident at this event, but the risks were high.

PUTTING IT ALL TOGETHER

Understanding your routes, understanding your areas, your crowd movement, and understanding your crowd profile, the people, and the demographics. Those are the elements of a RAMP analysis. In Chapter 16, and

[13] https://open.lib.umn.edu/sociology/chapter/21-1-types-of-collective-behavior/
[14] https://www.bbc.co.uk/news/uk-england-birmingham-47013978

Figure 5.8 Sydney Olympics control room.

the associated supplemental material on the publisher's website, we highlight students' case studies applying these models to their events.

We have applied these tools and techniques for over three decades for crowd safety planning for major events. We applied these tools for the Sydney Olympics, and we developed a crowd condition monitor (Figure 5.8).

We designed the system, and John Webster[15] coded it into Excel. The central display for the Sydney Olympics was the crowd condition monitor. It showed the control room staff the expected crowd density on the site, alongside the CCTV images of the current crowd density. There were crowd management options for regulating upstream crowd flow to prevent overcrowding at key locations. It was used to manage the movements of 11 million people over 14 days, taking the game schedule, taking the transport system, and incorporating it into the control room model.

The team had access to weather radar information, transport capacity data, train schedules, site models, and extensive CCTV. They could see any part of the site and the model is informing the team what they should expect to see. Early warning of congestion, at any part of the site, meant that congestion could be managed by upstream, hold and release of the crowd. Holding the crowd back for a few minutes, then release the crowd for a few minutes, repeat until the control rooms give the OK to resume message.

Behind these control room systems, we have used sophisticated tools for site modelling, flow modelling, crowd management strategies, and we assess them in the modelling framework. The control room only sees the "hold and release" instructions, using the KISS[16] principles.

15 https://www.linkedin.com/in/john-webster-08722468/
16 https://en.wikipedia.org/wiki/KISS_principle

Chapter 6

Crowd counting

> When you have mastered numbers, you will in fact no longer be reading numbers, any more than you read words when reading books. You will be reading meanings.
>
> W. E. B. Du Bois

Over the years, we have conducted dozens of major event crowd counts worldwide, on-site, in control room environments, or remotely.

During the MSc course, we performed counts for a variety of events, from the Royal Wedding (UK, Harry, and Megan, 2018), the Philadelphia Eagles Victory Parade (2018), Donald Trump's Inauguration (2017), the Anti-Brexit March (2019) to name a few. These counts are a team effort, with multiple inputs (using transport, car parking, flow rate calculations, CCTV/Video analysis) and cross-checking each other's work. We have had access to multiple on-site photographers, videographers and local TV/Radio/police helicopters. Crowd-counting is very controversial. Politicians always like to think they had many, many, many more people than they did, and organisers also want to inflate them for marketing.

THREE MILLION PEOPLE

Shortly after his election at USA president, Donald Trump announced there would be three million people attending his inauguration in January. In our work for the Ministry of Hajj (Saudi Arabia) the crowds amassed to three million pilgrims. We know what three million people look like. We know the area that three million people require for mass gatherings. We also know that certain politicians are prone to exaggeration.

The story is one we cover in our courses, so that we will only outline the basics here. If you want to hear the entire lecture on this interesting and funny story[1], it is in the *Introduction to crowd science* course.

[1] https://www.timeshighereducation.com/blog/i-am-scholar-caught-trump-inauguration-crowd-controversy

DOI: 10.1201/9781351053068-6

AREA CHECK

First, we always check available areas. If we use an initial estimate of two people per square metre and an expected three million, then there need to be at least 1,500,000 square metres of viewing space.

That is a square 1.23 km along each edge, full of people.

You should, of course, provide barriers, buffer spaces, emergency routes, food and beverage, concessions, PA (Public Address system) and large screens for viewing. In effect, you would need a larger area for the infrastructure. Also, the transport system would need to be able to move three million in the few hours to and from the event, and you would need to process/ticket check/security screen everyone.

The area on the Washington Mall is 1,100 metres long (the covered area from seating space to media tents) and 200 metres wide. A total area of 220,000 square metres, if this were full (to a safe capacity, based on the international guidance of 2 people per square metre), that is 440,000 people, far short of the 3,000,000 expected.

NOT 1.8 MILLION

The apparent record crowd attending an inauguration, during the 2009, Obama Inauguration crowd count is 1.8 million people. This proved to be inaccurate. Transport data (from Washington Metro) showed for Obama's 2009 inauguration 513,000 ridership (this equates to people using the metro to go to the event). Various media reports indicated a large walk-up, as the metro was at capacity. Also, the car parks were full, and 3,000 busses were in transit. This did not account for a missing million attendees. However, this was not an important consideration as we were performing a comparative analysis. Compared to the Obama crowd, how big was Trump's crowd? More, the same, or less? The actual numbers only mattered if the two crowds were of comparable size.

The Trump inauguration showed 193,000 D.C. Metro ridership for the same period, the car parks were 1/3rd full, and there were only 393 busses in transit on the day. We had high-resolution images (satellite) from Obama's inauguration, and this showed an occupied area of 175,000 square metres as the crowds extended well beyond the mall and around the Washington Monument. We could see the occupied area of the Trump Inauguration crowd covered only 61,000 square metres from photographs and video footage. One other key point was the live video footage and the arrival profile (the M in RAMP Analysis). The crowds were arriving at a steady flow rate throughout, and there was no observed exponential curve (higher crowd flow) as you might expect for a typical mass gathering.

1/3RD OF THE OBAMA CROWD

1/3rd of the occupied area (Obama's 175,000 square metres compared to Trump's 61,000 square metres), 1/3rd of the number of busses (3,000[2] for Obama, 393 for Trump), 37% of the metro ridership (Obama 513,000, Trump 193,000). You can draw your conclusions from these data. We also conducted a real-time crowd count for the Women's March on the following day (Jan 21st, 2017), which was three times bigger than the Trump Crowd, again that is an analysis and data we share on our lectures and courses.

References

Bump, Philip (January 23, 2017). "There's no evidence that Trump's inauguration was the most-watched in history. Period". The Washington Post. The Times spoke with crowd estimation experts who pegged the number in attendance on Friday at one-third of Obama's 2009 audience.

Frostenson, Sarah (January 24, 2017). "A crowd scientist says Trump's inauguration attendance was pretty average". Vox. "To try to settle the question of how many people attended the inauguration ceremony on the National Mall in Washington, DC, we reached out to Keith Still, a professor of crowd science at Manchester Metropolitan University" in the UK. (He analysed aerial shots of the crowds of both President Obama's 2009 inaugural address and President Trump's for The New York Times.) His conclusion is that the crowd on the Mall on Friday was roughly one-third the size of President Obama's".

HONG KONG

In 2019, Simon Scarr (Reuters) asked if we could outline the principle involved in crowd counting for an article[3] he was writing for the Hong Kong Protest march. The event organiser was claiming million people. He wanted to know about the methodology because they had the organiser estimate of numbers, the police estimate, and an estimate from Hong Kong University. The organiser's claims were five times the police and University estimates. The writer was really interested in the methodology, and to make accurate estimates from the event, they asked us *how to crowd count* so that they could make their own estimates.

---

[2] Source: Washington, DC Officials via the Washington Post.

[3] https://graphics.reuters.com/HONGKONG-EXTRADITION-PROTESTS/0100B01001H/index.html

AREA ANALYSIS

Using Google Earth Pro, it is possible to measure the route the protest would take. The graphics and calculations appear in the Reuters article, 145,000 square metres. Unlike the Inauguration crowds, this is a moving event, over 8 hours, of people marching through the area. An area analysis provides a maximum, at any given time, number of people on the route. This is a capacity that must be matched by the ingress and egress capacity. People need to come from somewhere at the start of the route and then go somewhere at the end of the route. This fact often eludes the crowd counting critics.

Over the last decade, we have used a polygon approach to measuring occupied areas. The Jacob's method[4] is a reliable approximation, but *a priori,* you can use sightlines to predict crowd build up, high-density sections and hence perform a more reliable risk assessment for an event. To do this, instead of mapping an area into grids, we use a set of triangles[5]. You can draft sightlines along parade routes, adding this to your RAMP Analysis and assess where you might expect high density. By coupling sightlines to your area/risk/congestion mapping, you can build an accurate assessment of risks. We teach this method in our online and classroom courses.

FLOW ANALYSIS

The transport capacity is often the limiting factor for these large events. Namely, the capacity of the station at the start of the protest and the capacity of the stations at the end of the march in the Hong Kong example. The entire MTR moves 4.9 million people per day[6]. So, in theory, the system had the capacity for one million protesters. Reuters recorded the assembly area for the entire event (8 hours), and we could assess capacity from that video footage. However, there is a first pass approximation for estimating the crowds. We can an a priori determine a rough-cut capacity analysis. The protest crowd will all pass through the narrowest point in the system. This was an area along Hennessy Road at the junction of Tin Lok Lane, some 20 metres wide. We can measure density in that area, typically around five people per square metre (this was a packed street and the flow rate, as shown in the Reuters video, was stop/start along the route), the flow rate was therefore around the 50 people per metre per minute limit. Our approximation, for 20 metres at 50 people per metre per minute (1,000 people per minute) times 60 minutes × 8 hours = 480,000 people passing through the narrowest point.

---

[4] https://en.wikipedia.org/wiki/Crowd_counting

[5] Any shape can be divided into triangles.

[6] https://www.statista.com/statistics/960380/hong-kong-average-daily-passenger-number-on-mtr-lines/

There was a second protest march two weeks later, with the organiser claiming two million people attending, but passing through the same narrowest point. The organisers claimed:

> We do a headcount by counting people one by one.
>
> *Bonnie Leung, the vice-convenor, told Reuters at a news conference.*

Counting ten people *per second* is one person counting 36,000 people per hour. That would take 55.5 person-hours of counting to count to two million. The event was only 8 hours, so if you had seven people, counting for 8 hours at ten people per second, no breaks, no double counting, you get to two million. You can assess the credibility of the claim for yourself, try and count to 100 in 10 seconds.

The Hong Kong Police count was 338,000, but they did not state how, but they did state they were measuring flow rates. Hong Kong University counted 500,000, using both density and flow sampling. Our first pass approximation, using density and flow was 480,000 people. You can see how the occupied area, the video analysis, the flow rates are all indicators of the numbers attending this event. When the organiser claimed two million attending the second event, which used more area, but still the crowd all flowed through a section. Reuters physically measured the area and recorded the flow rates. They sent the video for analysis. You can see the analysis, and video, in the Reuters article.

WHY IS CROWD COUNTING IMPORTANT?

The organisers claimed four times more than the other counts. Why should that be a problem? During the start of the Hong Kong protest, that there was a mass gathering of Chinese forces. Reports of 20,000 uniform police officers[7,8,9] in response to the claims of two million protesters. By overestimating the numbers attending the protest, there is a risk of a disproportionate response, which creates its own problems[10] and issues. We have seen similar incidents, a report of tens of thousands, on-street protests resulting in a disproportionate police response in Ireland, the UK, and across Europe, with the inevitable headlines.

7 https://news.sky.com/story/satellite-images-show-troop-build-up-on-hong-kong-border-11785250

8 https://www.theguardian.com/world/2019/aug/14/satellite-photos-show-chinese-armoured-vehicles-border-hong-kong-shenzhen-sports-stadium

9 https://www.smh.com.au/world/asia/thousands-of-chinese-police-hold-mass-gathering-next-to-hong-kong-20190731-p52clb.html

10 https://hongkongfp.com/2020/11/19/hong-kong-police-played-pivotal-role-in-radicalising-protests-in-build-up-to-poly-u-seige-policing-expert-report-says/

Logan P Kennedy in his thesis, Policing Protests: An Exploratory Analysis of Crowd Management Policies highlights issues relating to policing protests in the USA.

> This study examines police policies governing protest management to identify current U.S. police practices. The International Association of Chiefs of Police (IACP) provides model policies to help police agencies become familiar with best practices and develop their own policies. The IACP's model policy on crowd management and control was used to identify tactics that represent best practice standards for protest management in the United States. Through a content analysis of policies from a sample of U.S. police agencies, this study assesses agency compliance with the IACP model policy on crowd management and control, as well as alignment with existing protest management strategies.

We recommend the work of Reicher[11], Stott[12], Drury[13] and Cocking[14] for further reading on the topic of policing.

ECONOMIC ISSUES WITH INFLATED CROWD COUNTING

Overinflating crowd numbers can have a disproportionate response as we outlined above. Also, overinflating numbers has an interesting perception associated with future budgets. If an organiser is claiming a million people attend an event, more than doubling the actual numbers attending. That is the figure the media may report. However, it is not the number used during planning and management of the attending crowds. The police and security forces will perform their own pre-site checks, using realistic estimates, area calculations and understanding crowd dynamics. They would plan and budget for the expected (realistic) number attending.

Consider an organiser claiming one million people attend an event and that is the number reported in the mass media. The crowd management plan for (say) 400,000 people attending will include deployment of sufficient police/security/stewards/ushers for 400,000. The cost will be ££$$ X.

For another local event, and different organiser states 100,000 will attend. How do the politicians view the costs? They can take the higher number (first organisers claim) and the budget (for 400,000 – accurate attendance) and scales down the budget by a factor of ten. It is pointless

---

[11] https://risweb.st-andrews.ac.uk/portal/en/persons/stephen-david-reicher(a0a908db-1bb8-4d5e-ab30-f47643e35169).html

[12] https://www.keele.ac.uk/psychology/people/cliffordstott/

[13] https://profiles.sussex.ac.uk/p92858-john-drury

[14] https://research.brighton.ac.uk/en/persons/chris-cocking

complaining then, that the numbers were wrong. You should have made that clear for the inflated claim.

The flawed budget evaluation, taking the higher (inflated) organiser's number of people attending against the actual budget spent (for a lower, more realistic number attending), equates to budget expectations for other crowd numbers. We have lost count of the number of times we see this same problem, where there is a reduced budget for policing major protests because of the inflated numbers from organisers of previous events.

COUNTING BACKLASH

Event organisers, and the people attending non-ticketed events, do not like their claims checked. On social media, we receive aggressive responses to our crowd counting. However, to date, no one has proven the inflated numbers they claim.

One example of this, and the social media backlash, was the 23rd of March 2019 protest march. The organiser claimed one million attended the first event and one and a half million attended the second event. The inflated claims of numbers attending give rise to even more inflated claims for the next event, then the one after that, and the one after that. With Alan Wilson – Australia and Andrew Seeds – UK, we examined the "People's March" on the 23rd of March 2019. There was a flurry of media interest, and Wired[15] ran an article, and we had the usual social media criticism, without evidence of numerical analysis to prove that there were "a million people attending" claims.

ANDREW SEEDS – COUNTING THE PEOPLE'S MARCH

About Andrew

Andrew[16] is Events Broadcast Engineer turned Safety Advisor who has worked on events and filming projects large and small. A strong believer in the utilisation of data driven approaches to risk management, he is currently studying his MSC in Crowd Safety & Risk Analysis with Manchester Metropolitan University. Generally seen on his bicycle both around event sites and for leisure.

Introduction

On the 23rd of March 2019, London saw a significant gathering/march of persons to demonstrate for a second EU referendum. As many persons were expected the media asked us to look to provide a crowd count for the march

[15] https://www.wired.co.uk/article/brexit-march-peoples-vote-crowd-size
[16] https://www.linkedin.com/in/andrewseeds/

based on available data. Three methods were used to provide an attendance value for the march.

Utilising three independent methods focusing on different elements of RAMP analysis. For the routes the public transport capacity was calculated for the march area to establish the maximum value the public transport system could accommodate. Movement was established by calculating the numbers of persons passing a fixed location along the route through the review of publicly available time-lapse footage. The area of the crowd was established using a Helicopter flyover of the march that was broadcast live by various media outlets.

All methods were largely independent of each other focusing on different unconnected factors (although density estimates were present in two methods) and each analysis makes clear that the organisers' declared estimate was "over 1 million people" was highly unlikely to be correct. The analysis by the team based on our evidence suggested that the true figure was approximately 312,000–400,000. This work was subsequently picked up by various news outlets, including Wired, The Daily Mail, Full Fact and Channel 4 news.

Routes

The primary method of transport used by the marchers was the tube and other forms of public transport. There were no reports of significant overcrowding on the tube or stations having to close, or turn to exit only, or similar operational changes, to deal with the numbers of persons. However, a maximum theoretical number of persons able to be delivered to the march by the tube network was calculated based on the following information:

Three Key Tube Stations severing the march are Marble Arch, Hyde Park Corner or Bond Street (Peoples Vote, 2019). The three stations are served by a total of the three tube lines (The central line serves both Bond Street and Marble Arch) these are the central line, jubilee line and Piccadilly line.

- The Jubilee Line has a per train capacity of 817 persons and can run a peak service of 31 trains per hour in each direction (TFL, 2019).
- The Central Line has a per train capacity of 892 persons and can run a peak service of 35 trains per hour in each direction (TFL, 2019).
- The Piccadilly line has a per train capacity of 684 persons and can run a peak service of 24 trains per hour in each direction (TFL, 2019).

The total theoretical capacity for each line can be found by the following formula: Number of Trains Per Hour Each Line × Per Train Capacity × 5 (Hours).

- Jubilee Line = 62 × 817 × 5 = 253,270
- Central Line = 70 × 892 × 5 = 312,200
- Piccadilly Line = 48 × 684 × 5 = 164,160

Also, the theoretical number of exits possible if the following factors were in place:

- All lines ran a full weekday peak service for the full 5-hour period.
- All trains were full to the maximum in both directions at all times.
- All passengers left at one of the three stations serving the march. (No passenger passed through a march serving station without getting off.)
- There were no delays on any of the lines at any point.

If the above factors were in place the maximum capacity would be
253,270+312,200+164,160=729,630 persons

TFL (Transport for London) states that it is impossible for them to run a peak level service sustained for more than an hour or so due to driver breaks/changeovers and line capacity. It is also deemed in probable that, if the above were true, this would have caused significant and widespread network congestion which was not reported.

It is also acknowledged that persons could arrive via Bus or Coach, approximately 200 coaches (Peoples Vote 2019) were organised for the march, which were likely to have a total capacity of 11,200 (based on 56-seater coaches). It is not quoted how many busses were full of marchers, however, the capacity for this would likely be limited to less than 1,000 actual busloads of people, around 86,000 persons based on "New Routemaster" Style Buses (Wikipedia, 2019)), however, it is not possible to verify this figure.

It would also be required to remove the normal Saturday Traffic for these stations which is as follows:

- Marble Arch – 15,796
- Hyde Park Corner – 7,400
- Bond Street – 45,840

Normal Saturday Capacity for the three stations is (15,796+7,400+45, 840)=69,036. The number of passengers, 69,036, would need removing from the 729,630 people. Walking and cycling and other modes of transport would be seen as insignificant given the low population density of Central London and the nature of the event.

Area

Using Google Earth Pro and associated press helicopter footage (available on YouTube,[17,18], 2019) it was possible to draw around the area of which the

[17] https://www.youtube.com/watch?v=u1QBYCycNCY
[18] https://www.youtube.com/watch?v=b-bTgR7B1TM

marchers were occupying. This was worked out at around 156,246 square metres. Using various video footages, including the BBC live coverage[19] footage, it was possible to estimate the density of the crowd during this flyover period. This was estimated by Alan Wilson (Australia) and G. Keith Still (UK) at around 2–2.5 persons per meters squared, giving a total value of around 312,000–400,000 persons. More in-depth analysis could have been done with specific sections of the march during the flyover; however, this was not possible given the time pressure of the group due to press deadlines, this would likely only serve to reduce the range of the estimate provided by this method.

Movements

As a march people were moving towards Westminster square; using publicly available footage (YouTube 2019). This video states that it covers just over 2 hours of marchers passing in 6 minutes and 34 seconds. For the purposes of this analysis, it is accepted that this video is 2 hours long. This will lead to a slightly higher flow rate within the figures. The video is 394 seconds long and represents 2 hours of footage (7,200 seconds). As such every second of video is 18 seconds in real life. The video's frame rate is 25 frames per second. This means that each frame represents 0.72 seconds. The video is low resolution; however, it is possible to track the movement of distinctive objects.

The basis of the flow rate analysis is as follows. Starting at 0:10 two points was picked at the edge of the tracking range and the number of frames it takes to pass through the defined distance counted. Record each point independently along with a decimal value for how far across the road they are (closest to the camera is 0 and the building is 1). This process is repeated after every 20 seconds of video.

There is a linear relationship between the distance to the camera and the number of frames it would take to pass through the same area. This is because the camera is a single point, and the two reference points are fixed. As such dividing the frame count by the position will give a weighted number of frames it would take to cross the same distance. The average of these two values will be used for further analysis. The number of frames is then timed by 0.72 to give the number of seconds it would have taken to travel the determined distance. Both values are then dividing by the distance in meters between the two reference points to give meters per second value.

Formula's used

- Frame Count 1 / Position 1 = Weighted Frame 1
- Frame Count 2 / Position 2 = Weighted Frame 2

[19] https://www.bbc.co.uk/news/uk-politics-47678763

- Average of Weighted Frame 1 & Weighted Frame 2 = Average Weighted Frame
- Distance / (Average Weighted Frame × Seconds Per Frame) = m/s of crowd
- Seconds Per Frame × Video Frame Rate × Seconds Between Time Stamps = Seconds of Real-Time per Time Stamp (Not Shown)
- Crowd Density × M/S of Crowd × Width of Road = Flow Rate (Seconds - not shown)
- Flow Rate × Seconds of Real-Time per time stamp = Number of persons passed through in timestamp (Tables 6.1 and 6.2)

This analysis suggested that 182,906 people had walked past the camera during a 2-hour period. Further analysis was then done to extrapolate this data across the total duration of the march (accounting for the fact that it would take time for the front of the march to reach the point of the camera).

Formula's used

- Distance From Start Point / (60 × Average M/S of Crowd) = Length of Time to Get to Camera
- Duration of March as advertised – Length of Time to Get to Camera = Available Marching Time at location
- (Total Number of Crowd Expected to have passed through during analysed section/Total Duration Analysed) × Available Marching Time at Location = Total Crowd Expected to Flow Through
- Duration of March (as advertised) 300 minutes
- Length of time to get to camera 39.78 minutes
- Available marching time at location 260 minutes
- Total crowd expected to flow through system 396,634

This gave a value of 399,634 persons walk past this point on the march throughout the day of the March.

Persons/profile

From the nature of the march and event, there was no knowledge of the specific demographic that turned up to the march; however, it would be expected that this was reflective of the overall UK population in terms of age and gender split.

Table 6.1 Crowd count fixed values

Seconds Per Frame	*Width of Road*	*Video Frame Rate*	*Seconds Between Time Stamps*	*Position of Camera in route*	*Distance Measured Between*
0.72	20.21	25	20	1200	13.13

Table 6.2 Crowd count measured values

Time Stamp	*Frame Count 1*	*Position1*	*Frame Count 2*	*Position2*	*Crowd Density P/m²*	*Weighted Frames 1*	*Weighted Frames 2*	*Average Weighted Frames*	*M/S of Crowd*	*Persons per 20 video[23] seconds*
10	12	0.5	24	0.9	2.5	24.00	26.67	25.33	0.72	13,093
30	22	0.5	41	0.9	2.5	44.00	45.56	44.78	0.41	7,408
50	48	0.8	7	0.1	2.5	60.00	70.00	65.00	0.28	5,103
70	35	0.8	8	0.2	2.5	43.75	40.00	41.88	0.44	7,921
90	23	0.5	33	0.9	2.5	46.00	36.67	41.33	0.44	8,025
110	50	0.6	10	0.3	2.5	83.33	33.33	58.33	0.31	5,686
130	22	0.7	10	0.4	2.5	31.43	25.00	28.21	0.65	11,756
150	23	0.7	4	0.1	2.5	32.86	40.00	36.43	0.50	9,105
170	10	0.4	43	1	2.5	25.00	43.00	34.00	0.54	9,756
190	30	0.7	4	0.2	2.5	42.86	20.00	31.43	0.58	10,554
210	19	0.5	12	0.3	2.5	38.00	40.00	39.00	0.47	8,505
230	9	0.3	21	0.7	2.5	30.00	30.00	30.00	0.61	11,057
250	8	0.3	27	0.8	2.5	26.67	33.75	30.21	0.60	10,980
270	11	0.2	25	0.9	2.5	55.00	27.78	41.39	0.44	8,014
290	28	0.9	16	0.3	2.5	31.11	53.33	42.22	0.43	7,856
310	58	0.9	21	0.2	2.5	64.44	105.00	84.72	0.22	3,915
330	15	0.3	27	0.6	2.5	50.00	45.00	47.50	0.38	6,983
350	31	0.7	21	0.6	2.5	44.29	35.00	39.64	0.46	8,367
370	5	0.2	15	0.9	2.5	25.00	16.67	20.83	0.88	15,921
390	15	0.7	6	0.2	2.5	21.43	30.00	25.71	0.71	12,899
									Total Number of Crowd Expected to have passed through during analysed section:	18,2906
									Average M/S of Crowd	0.502792
									Total Duration Analysed:	120

[23] 6 minutes real time.

SUMMARY

Within the above RAMP analysis there was significant information available to assess the Areas and Movements of the people involved in the march; these two figures correlate to each other strongly (with the top end of the Areas Estimate being reflective of the Movements estimate); while the routes available to them had significantly higher capacity than otherwise estimated this was based on idealistic scenarios which would be near impossible to be truly reflective, as such it was felt that this figure could be discarded.

REFERENCES

BBC News. (2019). Million joined Brexit protest, organisers say. [online] Available at: https://www.bbc.co.uk/news/uk-politics-47678763 [Accessed 16 Aug. 2019].

Carpen, A. (2019). Timelapse part 2 of Peoples Vote March - London. 23 March 2019. [online] YouTube. Available at: https://www.youtube.com/watch?v=sIVbVt6K7OQ [Accessed 16 Aug. 2019].

Churchill, D. (2019). Tusk Taunts Britain. [online] Pressreader.com. Available at: https://www.pressreader.com/uk/daily-mail/20190328/281814285215978 [Accessed 16 Aug. 2019].

Craig, Z. (2019). 16 Things You Might Not Know About London's Buses. [online] Londonist. Available at: https://londonist.com/2016/09/london-bus-facts [Accessed 16 Aug. 2019].

Made by TfL. (2019). Tube trivia and facts. [online] Available at: https://tfl.gov.uk/corporate/about-tfl/what-we-do/london-underground/rolling-stock [Accessed 16 Aug. 2019].

Marsh, S. and Rubin, G. (2019). Brexit march: '1 million' rally for people's vote - as it happened. [online] the Guardian. Available at: https://www.theguardian.com/politics/live/2019/mar/23/brexit-hundreds-of-thousands-expected-to-march-for-peoples-vote-london-live-updates?page=with:block-5c9650d4e4b0a5422e63a3fc [Accessed 16 Aug. 2019].

People's Vote. (2019). Put it to the People - Frequently Asked Questions. [online] Available at: https://www.peoples-vote.uk/frequently_asked_questions [Accessed 16 Aug. 2019].

People's Vote. (2019). Put It To The People March estimated to have broken the 1 million people mark. [online] Available at: https://www.peoples-vote.uk/put_it_to_the_people_march_estimated_to_have_broken_the_1_million_people_mark [Accessed 16 Aug. 2019].

Stokel-Walker, C. (2019). We counted the people on the Brexit march, so you don't have to. [online] Wired.co.uk. Available at: https://www.wired.co.uk/article/brexit-march-peoples-vote-crowd-size [Accessed 16 Aug. 2019].

The Independent. (2019). '1 million' Put It To The People protesters stage historic rally for a second referendum. [online] Available at: https://www.independent.co.uk/news/uk/politics/brexit-march-london-put-it-to-the-people-live-updates-article-50-a8836386.html [Accessed 16 Aug. 2019].

Transport for London. (2019). Underground services performance. [online] Available at: https://tfl.gov.uk/corporate/publications-and-reports/underground-services-performance [Accessed 16 Aug. 2019].

Urwin, R. (2019). A million march in London for People's Vote as the cabinet plots a putsch. [online] Thetimes.co.uk. Available at: https://www.thetimes.co.uk/article/a-million-march-in-london-for-peoples-vote-as-the-cabinet-plots-a-putsch-qp57ntw6g [Accessed 16 Aug. 2019].

Wilson, A. and Still, G. (2019). Introduction to Crowd Safety and Risk Analysis. [online] Gkstill.com. Available at: http://www.gkstill.com/Education/mod/forum/discuss.php?d=113&mode=3 [Accessed 16 Aug. 2019].

CONCLUSIONS – CROWD-COUNTING

If we know the narrowest point in the system, and the duration of the event, we can make an estimate of the crowd numbers using a flow calculation. We can do this a priori, to estimate expected numbers. If we know the transport capacity[20], then we know the upper limits for numbers that might attend a non-ticketed event, like a protest. If the event is non-ticketed, and mostly static in nature (spectating events) then we can estimate numbers, and measure ingress flow, to manage the event, before it reaches capacity, using area analysis.

The methodology of crowd counting, of flow monitoring and real-time decision support, using visual techniques to assess crowd sizes, is central to our online courses and for the MSc. We provide templates and tutorial on real-time applications. Our MSc students and short course delegates have applied the models to their events, around the world.

[20] In one, heated social media exchange, we highlighted that the crowd stopped at the end of the march and that the stations entrances were visible, but no queueing was observed. It was not free flowing for the duration, they moved along the route, then stopped at the end.

THE FINAL WORD

Crowd counting always generates a stream of criticism and negative social media[21] and hate email. We do not crowd count to win friends.

The Trump[22] count and the Eagles Victory parade count generated a significant negative feedback, choice words, profanity, and a few threats. However, to date, no one has supported their inflated number with evidence. You can see our methodology above, and we encourage you to check your own sites/events.

[21] Except the security company and police who planned and managed the event. Their estimate, *a priori,* and confirmed on the day, was 440,000 maximum attending.

[22] https://www.vox.com/policy-and-politics/2017/1/24/14354036/crowds-presidential-inaugurations-trump-average

Chapter 7

Control room information

> Having knowledge but lacking the power to express it clearly is no better than never having any ideas at all.
>
> Pericles

We discussed real-time counting in the previous chapter, and we also outlined the use of 3D graphics in a control room setting. You can use these for both monitoring the crowd density and flow and as part of your audit trail. For example, at 3 pm area. It was between two and three people per square metre and free-flowing.

These audit trails can become an asset for two reasons. First, they build up your knowledge of the crowd dynamics, hence developing risks around your site. Second, they allow others to examine your operations plan. The latter can be valuable evidence of your care and attention to crowd safety. Accidents can happen and demonstrating your records can be vital evidence of your safety practices, specifically in lawsuits against you or your organisation.

We developed these techniques from two different perspectives. First, you will be able to identify problems, specifically flow/density issues, before they develop into difficult to manage situations. Second, as the expert witness will examine your operational process, procedures, and documentation, these audit trails are clearly demonstrating you record, monitor, and can act early if a situation begins to develop.

IS IT POSSIBLE TO CONTROL ATTENTION?

The key to using this information is data positioning and our mind's ability to recognise problems. We need to consider not only what information we use but also where you locate it in your control room.

DOI: 10.1201/9781351053068-7

To set the scene, please view this Ted talk by Apollo Robbins.

https://www.ted.com/talks/apollo_robbins_the_art_of_misdirection?utm_campaign=tedspread&utm_medium=referral&utm_source=tedcomshare

You can google "Ted Talk Apollo", the title of the talk is "the art of misdirection".

WHY THIS IS IMPORTANT WHEN WORKING IN CONTROL ROOMS

In previous chapters, we talked about perceptions and how they shape our view of the world. In the TED talk, Apollo Robbins is manipulating the audience's attention; he understands how people process information, what levels of information they will miss[1], how much they will process, and how much they will remember. This is the magician managing the audiences' perceptions of reality, directing the audience's attention, and understanding how much (or little) we are capable of processing at any single moment.

His graphic of "Frank" turning round to search through the filing cabinet perfectly illustrates the need to keep information in full view; hence, the development of the real-time graphics for the Sydney Olympics and other sites. Information, specifically visuals of crowd density presented in an easy-to-understand format (an image), kept in full view, help you to evaluate the crowd dynamics.

WHY DO WE NEED RISK MAPS?

That is why we need crowd risk models? That is why we use graphics (faster for the eye to process, contains more information, everyone can see the same information at the same time), to keep relevant information in front of you. So that when you work in a control room environment, you are not relying on, or referring to your memory, or consulting with your colleagues and friends around you. The information is there in front of you.

It is vital to understand why, for large events like the Sydney Olympics, this helps us understand problems that are starting to arise while we still have time to do something about them. This is the reason we called it "red-flagging" in the DIM-ICE section of this book. Suppose we understand what we are expecting if we can see that things are not going to that plan. Then, we can do something about it.

---

[1] Did you spot when he changed his clothes? Go back and watch the clip again.

Figure 7.1 Using graphics to assess crowd density. (Graspop Metal Festival 2018, Dessel, Belgium, Picture used with permission from Donald Goedheid.)

EARLY DETECTION

If something starts to go wrong, and you fail to recognise it early enough to intervene, you might not deal with the situation if you make your decisions too late. You only need to review past disasters to realise the impact of late decision-making, failing to understand what might happen next and/or failing to understand the consequences of a situation evolving. These graphics help your early recognition of potential crowd problems.

The use of simple graphics in front of the operator and crowd management team, or if you provide stewards/ushers with graphics to identify high density/flow and alert their supervisor, these systems provide insights to the situation as they evolve. For example, the image below shows an operator with his printouts assisting his decision making. When the density from this angle looks like this, shut the gate (Figure 7.1).

He has taped images of different crowd densities from the same camera view. He is monitoring the crowd density as it builds up and informing the ground staff of the situation. He is also recording the crowd build-up over time. 2 pm density=1–2 people per square metre, 2:15=2–3 people per square metre. He clearly indicates when the density might reach a critical level and act before the situation develops into a crush. Therefore, when the crowd density reaches his cut-off level[2], he is able to act upon it. He is not trying to understand, interpret or remember things. He can look at his

[2] 0%–50% = Green, at 50% amber – get ready, 50%–80% amber and at 80% (red) control crowd flow.

graphics and very quickly cross-reference. When it looks like that, we need to implement this management control or stop the crowds moving or use a hold and release mechanism. Use these graphics to plan and manage the crowds on his site, with a crowd management plan in real time.

As Apollo Robbins demonstrated in the Ted Talk, these tools keep your attention focused on the situation, without diverting your mind from the problem.

FREE DOWNLOAD

We have produced a series of graphics for your use, the first set are in metres,

You can also download these from this link

https://crowdrisks.com/crowd-graphics.html

Both people per square yard and square yards per person references (the USA highways capacity manual uses area per person). If you want something specific for a site, send me an email (GKStill@me.com), we are always happy to build these kinds of models with you for your site use, using the software that we developed.

Chapter 8

Queueing systems

> Airports drive me mad. I don't mind the flying; it's all the hassle before you get on the plane and afterwards, including walking five miles through corridors to the point where you queue for ages to check passports and hope your luggage has arrived safely.
>
> **June Whitfield**

A queue will form when the number of people arriving at a location exceeds the rate at which you can process those arrivals. The difference between people arriving (A) and people departing (D) over a period (T) is the number of people waiting in line (N). Queues form when the number of people who arrive exceeds the rate at which you can process them through or into the system. It could be a ticket check, security screening, passport checking, Covid-19 screening, anything where people need to pause for a process.

Queuing Systems can become quite complicated to assess if there are different processes involved. For example, queue jumping, or switching queues, due to slow operators, problems in the process (operator or mechanical issues).

You should plan for both efficiency and operational cost of queueing before the event starts. It may come as no surprise that there is a branch of mathematics dedicated to estimating queues, Queueing Theory[1].

ARRIVAL RATE AND SERVICE RATE

There are two primary values required to evaluate queuing systems. First, the arrival rate (the number of people arriving over time) and, second, the service rate (the number of people you can process over time). You can use these two values to determine a wide range of useful parameters, such as the utilisation rate (operational efficiency), cost (number of service points) and waiting times. However, mathematics alone does not address the human factors involved in queues.

[1] https://web.mst.edu/~gosavia/queuing_formulas.pdf

DOI: 10.1201/9781351053068-8

IT IS MORE THAN JUST THE MATHEMATICS

Historically, the mathematics of queueing theory evolved from process engineering, such as manufacturing. You estimate the time to process each stage of a manufacturing process, the number of components you need to ensure efficient operations, the stock and the operational costs.

Before developing crowd science, we worked in manufacturing, as an operations manager, and then as a trouble-shooter/consultant for manufacturing processes at a wide range of industries. Applying queueing theory, specifically the cost function analysis, was an essential tool of the trade used to analyse and streamline the various manufacturing and stock holding processes.

The bulk of the queueing idea and its associated mathematics fail to address the systems where *people* are waiting in line. Parts do not change their mind or get frustrated in queues; people do.

When we began designing and teaching the principles and applications of the crowd queuing system, there were two problems to overcome. The first of these was to simplify the complexity of queueing theory into a teachable and usable format. The second was to teach the basics of a "first pass approximation" or a "rough-cut analysis".

Simulations and models can provide numerical values for "time in a queue", and we have used a wide range of queueing formulas for crowd projects and site design. These are sophisticated programs, but they allow the user to calculate the physical parameters associated with queuing. However, human behaviour, such as the frustrations that can build up in long queues, is not part of queueing theory's mathematics.

The human factors may include queue-jumping, operator fatigue, changes in the environment, bright sunshine, sudden downpour. These can affect how people feel and behave in queues. There is a need to understand the human factors as well as the mathematics. Let us start with the system you need to evaluate when designing a queueing system, such as an entry gate, with a process, such as security, Covid-19 screening, ticket checks.

You can estimate the arrival rate from the transport system capacity. Look at how and where people are moving towards your entry point. What at the frequency and capacity of the trains, cars of busses? How many people are on those modes of transport? Where are the pubs and clubs, fan zones? When will the people from those areas walk towards the entry point? For this, use the RAMP Analysis (outlined in previous sections of this book). Estimate the maximum rate at which people may move towards the entry point. Then estimate how that movement over time may look (the M of the RAMP Analysis). Secondly, you may have a measure of the time it takes to process any single individual (an average is useful here), be it 20 seconds or 15 minutes[2].

[2] At the time of writing this, passport control operators are reporting that covid checks, the inevitable wait for results, passport checks are taking considerably longer and hence there are significantly larger and longer queueing and waiting times.

As an example, if an airport has 5 aircraft arrive per hour, and each carry (say) 80 passengers. Then the arrival rate is 5×80=400 passengers. Let us assume we have 20 members of staff in passport control.

400 divided by 20 (the number of staff)=20 passengers per hour per member of staff. Divide that by 60=3 people per minute per member of staff.

However, if the system changes and it now takes 15 minutes to Covid check and passport check each passenger. That is 4 people per hour per member of staff, or 80 passengers per hour for all 20 members of staff. To clear four hundred passengers will take 5 hours. You can see from this example why using a simple estimate can forewarn serious delays at this example passport control.

THREE CONDITIONS TO UNDERSTAND

There are three conditions to understand when estimating crowd queues. The first is when the arrival rate is greater than the service rate when the demand exceeds the capacity. Queues will form and, if there is limited space, then the queue may develop into a high-density, high-risk environment.

When large event crowds move towards trains, railway platforms can become overcrowded. If left unmanaged, this can be a considerable risk to life and limb. If you google "overcrowded Indian train station", you will find dozens of clips of high-density crowds packed on platforms and overflowing the stations. The transport demand exceeds the transport capacity, overcrowding results from high-density queueing, leading to disasters[3] in the Indian transport system.

Regulating crowd flow onto the platform is critical to balancing safety and security, keeping crowded spaces safe. When we teach in India, this problem highlights the issues of crowd management across the continent. Understanding these risks and putting inappropriate crowd management solutions is a core lesson of our Indian Police Training Course.

Overcrowded spaces can lead to frustration in individuals, and the crowds' adverse reactions/bad behaviour can lead to public order issues. Overcrowded platforms can force people to fall onto the tracks. Hence, crowd queueing will be a critical safety calculation in environments to minimise and manage the risks associated with overcrowding.

There is a secondary risk due to overcrowding. Queueing areas become soft targets for crime, disorder, or a malicious attack on crowds. Queueing models are important to avoid levels of frustration, to prevent creating soft targets and to avoid problems in your crowd management. Essentially, get your queueing systems right and you have a more efficient process, a calmer, less frustrated crowd, and that will impact their spend, enjoyment and desire to come back and spend more.

[3] https://en.wikipedia.org/wiki/List_of_railway_accidents_and_incidents_in_India

ARRIVAL RATE IS LESS THAN SERVICE RATE

The second condition is when the arrival rate is less than the service rate. Where you can process more people per hour than you expect to arrive. There will be no queueing, but your operating costs increase.

In our previous example of the four hundred passengers arriving per hour and we want to have no queues. We know it may take 15 minutes per person, so we can process four people per hour per operator. We would need one hundred staff to clear that arrival rate.

If we do not have space or money to facilitate one hundred passport booths, we can estimate how many booths we can facilitate, then how long it will take to clear that queue. Obviously, if more aircraft arrive in the next hour, the queue from the first hour will overlap the queue from the second hour, and so it goes on and on. If you have experienced passport control in the months following the resumption of flights, you will know the problem first-hand.

BALANCED SYSTEMS

This leads us to our third condition and the common error observed in overcrowded queueing systems. What happens when the arrival rate equals our service rate? When we run workshops, we stop at this point and allow the delegates to discuss the dynamics of an apparently balanced system.

In summary, we have three arrival rate and service rate conditions. These are:

Arrival rate is greater than the service rate, a queue will form
Service rate is greater than the arrival rate, it is not cost effective
Arrival rate equals the service rate, what happens

If your arrival rate equals your service rate, it should work? The problem is that the apparent balanced system assumes everything works perfectly. That the arrival rate is constant, and the service rate is also constant.

Things are never that simple. People arrive in clusters, for example, trainloads at a time. Some people can take a lot longer to process, to get served, to check, than others. If you cast your mind back to the last time you were in a supermarket queue or bag check at an airport, we have all experienced that situation where the person immediately before you are taking an EXTRAORDINARILY long time to complete their transaction/process. It could be the request for longer legroom (and the inevitable check for seating options) or a problem with vouchers, finding the right change, digging through pockets or handbags for credit cards, or deciding to have a chat about the weather, their cat, or .... There is an endless list of potential delays due to these human factors. That is where the main difference between the mathematics of queueing theory for manufacturing and the human factors involved in crowds queueing.

If you try and balance your service rate to your expected arrival rate, then your system will fail due to these human factors. It could be traffic delays, three busses arriving at once. It could be equipment failure, operator fatigue, a forged ticket, a suspect shape seen in an X-ray. Arrival rates and services rates are never constant where people are involved.

HOW DO WE "BALANCE" A SYSTEM?

You need to provide a contingency for the above. How do you do that? You can estimate your average arrival rate and average service rate for any period. Then you must provide additional resources for the service rate because things can and will go wrong in real life. Ticket machines can fail to operate, communication networks can run slow, or fail, a bunch of tickets might be damaged or be wet and difficult to scan. There might be problems in reading or scanning barcodes on wristbands due to incorrect application. We once heard of an event that had a very efficient search and ticket checking operation. They knew their service rates, their arrival rates and had provided sufficient, trained operators. Everything worked well until a coach load of supporters arrived, all dressed as bananas. How do you search a large foam for costume? It can create a delay in the entry process, and, on that day, the entire entry system was slower due to that coach load of bananas. You need to add a contingency factor; 20% is a good first pass approximation. Work out your arrival rate, your service rate and add 20% to your service capacity for contingency.

MOVEMENT OVER TIME

However, to evaluate queueing, you need to understand the crowd arrival profile, the movement of people over time, the M in the RAMP analysis. In other words, if you are expecting late arrivals, all our demand is going to close to the event start. If you are expecting early arrivals, all the demand will be 2 or 3 hours before the event starts. In Figure 8.1, we can see the three types of arrival, early (slopes down), late (slopes up) and the trainload (up, down, up, down) profile – the M in the RAMP analysis outlined this in more detail. If we consider an anticipated late arrival as an example (Figure 8.1), we can estimate the curve, then in each section, if we add 20% additional resources, we can minimise the potential queueing.

If you do not know your service rate, then measure it, count the number of people per period (15 minutes, for example). By counting/measuring your arrival/service rate for a few events, you can build up a profile for your site.

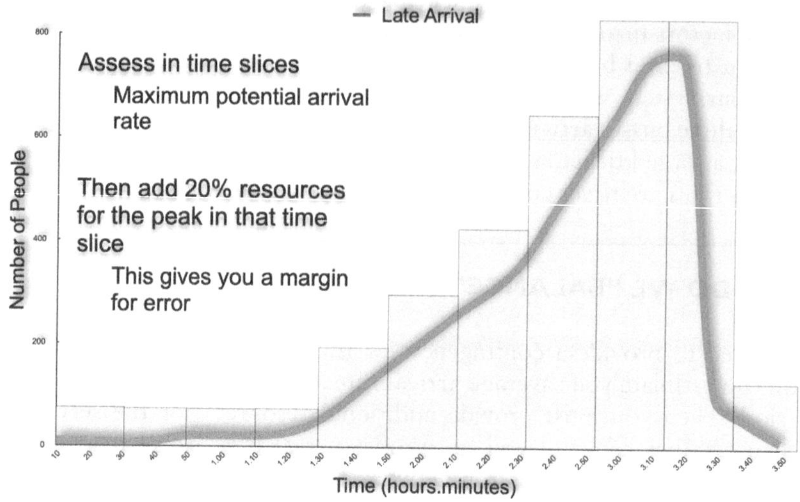

Figure 8.1 20% capacity over expected arrival demand.

You can then make reasonable estimates for future events. You can create an estimate of the shape of your arrival profile curve. Add 20% to your estimated service rate. That is your first pass approximate; measure again and check you have effectively balanced your system.

You can evaluate queueing systems and you can provide contingency (resources, spaces) using a simple rough-cut estimate. Whether you are calculating a flowing system or a queueing system, consider an example, narrow section on a route to an event space. Consider a narrow section on a route towards an event space. Let's say it is 10 metres wide (approx. 33 feet). If we know that there may be 145,000 people trying to move along a route at the peak hour (say just before an event starts), then we can work out system dynamics in advance.

A 10 metres (approx. 33 feet) wide road, at 82 people per metre per minute (25 people per foot per minute), would have a flow rate of 820 people per minute, or 49,200 people per hour. If we are expecting 145,000 people in that hour, then the system will fail as the width will only service 49,200 people per hour. The demand (145,000) exceeds the capacity (49,200) a queue will develop, and if we have limited space, there will be a significant risk to life and limb. It is not rocket science. Simple mathematics on a spreadsheet will show you there is a problem, and you can change something in the system to avoid the risks of overcrowding. That is the essence of working out queuing formulas, which is how to plan safe passage for your event.

Another example, from Sydney Olympics, which was expecting 500,000 visitors per day. They had limited space, two ways flows, and they added a bridge, which solved the basic problem relating to circulation and crowd flow. This allows the system to work as a one-way system if we consider one section of that system. There are three sources of crowd flow that combine into one. So, three routes leading to one route. We expected 30,000 an hour coming from the station, 40,000 an hour from the bridge and 10,000 an hour from the pavilion. The problem is how wide should the combined road be to accommodate these three flow rates. The total is an anticipated 80,000 people per hour. We know that our optimal flow rate is 82 people per meter per minute (25 people per foot per minute). Put the book down, or cover up the rest of this page, and try and work it out, before reading on.

Hopefully, you will have tried that little exercise and produced a figure. Let us run through the numbers and calculation with you. 30,000 per hour from the station 40,000 per hour over the bridge and 10,000 per hour from the pavilion. So that is 80,000 people per hour. That is the anticipated maximum flow, or the potential demand, on the system. We need to make that route wide enough to accommodate that demand/flow.

80,000 per hour divide that by 60 (to get people per minute) and we get 1,333.3 people per minute. Remember, we want to do our calculations in people per metre (or per foot) per minute. We know that the value of 82 people per metre per minute (or 25 people per foot per minute) is the maximum flow rate we might expect. If we divide 1,333.3 by 82, we get 16.26 metres (1,333.3 people per minute divided by 25 people per foot per minute=53 feet). We know what the demand is, and now we have worked out the required width of the road.

Then, of course, we need to add 20% for contingency, which comes to 19.5 meters (64 feet) but remember they build these roads with big yellow bulldozers. So, let us call it 20 metres (70 feet). We have just designed a wide enough system to cope with the anticipated demand and built a contingency into our calculations. And that is it.

It was interesting to note that it took over a year of simulation analysis and several thousands of dollars to produce the same figure. That is the simplicity of a rough-cut analysis, and it can provide a quick order of magnitude for both a crowd flow analysis and a queueing system.

THE FUNDAMENTAL QUEUING FORMULA

We have used sophisticated simulations for both flowing and queueing systems. There are different variations of the queueing formula (around 128 different classifications). We have also designed systems for queueing around the world. In every case, we use the first pass approximation is our initial estimation formula. It looks like this:

$$N = T \times (A - D)$$

where T=Time period (minutes or hours), A=Arrivals in period (T), D=Departures in period (T), N=number of people in the queue (after-time T).

Figure 8.2 Queueing diagram.

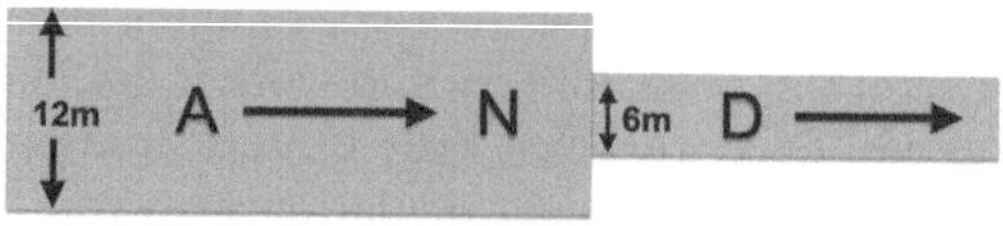

Figure 8.3 Flowing system.

As you can see, departures are the same as the service rate. How many people pass through this part of the system over time. With this, we can estimate the space, time, and stewarding required to manage the system. Here are a couple of examples. First, a queueing system, where people arrive, a process is involved, then they depart. In this example, N is the area we need to estimate for our queue (Figure 8.2).

If, say, we are expecting 500 people to arrive per hour, and we can process (ticket and bag check) 300 people per hour. After 3 hours, how many people might you expect in the queue? 1,500 arrived, 900 were processed, the difference is 600 people. So, you might expect N (the queueing area) to have 600 people in that space. From our chapter on crowd dynamics, we know we should estimate four people per square metre[4]. 600 people at four people per square metre you will need, 150 square metres to accommodate that queue. And if you do not have that space, then you may have a crowded space, you may develop a risk of injury to the crowd. If this were, say, a platform, and you anticipated 600 people to accumulate but only had 100 square metres on the platform, then you might expect there to be six people per square metre. That would not be a safe queueing area, specifically with trains arriving. Try working out some different numbers for arrivals and departures, for the queue build up, for the space available. Perhaps create a spreadsheet for this. You will see that if you know your arrival rate and your service rate (departure rate), you can evaluate your space and reduce potential overcrowding risks.

You can use the same formula for flow rate analysis. If we consider a 12-metre-wide (13 yards) road as indicated in the diagram below, that then narrows to 6 metres (6 ½ yards) wide (Figure 8.3).

We can use flow rates to work out the queuing issues. As outlined in the chapter on Crowd Dynamics, the crowd will flow at 82 people per metre per

[4] Four people per square metre, pre-Covid. Two square metres per person during the covid pandemic. If we factor the social distancing, wearing facemasks, levels of ventilation, vaccination, etc. etc. the density may change, but the formula is the same.

minute (25 people per foot per minute). 82 times 12 metres would be 984 people per minute, moving from A to N. Then the road gets narrower, it goes down to 6 meters. Six metres times 82 people per metre per minute will allow 492 people per minute through that section. The same results apply if you calculate in yards, 12 metres=39.37 feet (approx.) 39.37×25=984.25.

There would be a maximum flow of 984 people per minute arriving at N, but only 492 departing N, every minute. There are 492 people that do not pass N that start to form a queue in one minute. In 10 minutes, there would be 4,920 people that did not pass N. Therefore, in 10 minutes, there need to be at least 1,230 square metres, to avoid overcrowding (4,920 divided by 4). If there are less than 1,230 square metres, there will be a problem of overcrowding. The number of people arriving at N exceeds the number of people that can pass N and move to the right. We can apply the same queueing model to a flow rate and a discreet system of arrivals and processing (ticket check, security check, bag check, covid check, passport check, etc.).

$$N = T \times (A - D)$$

In both systems, when the arrivals exceed the departures (arrival rate exceeds the service rate), there will be a queue. Of course, this is an oversimplification of this kind of system, but that is the objective of a first pass approximation. If the system fails this kind of analysis, then it will fail when you reach those conditions. If it fails a rough cut – it fails.

Now let us look at a system with two elements, a ticket check (process) and a flow rate. Consider the system in the diagram below. To the right, there are six 1 metre wide gates. At each gate there is a ticket checker (Figure 8.4).

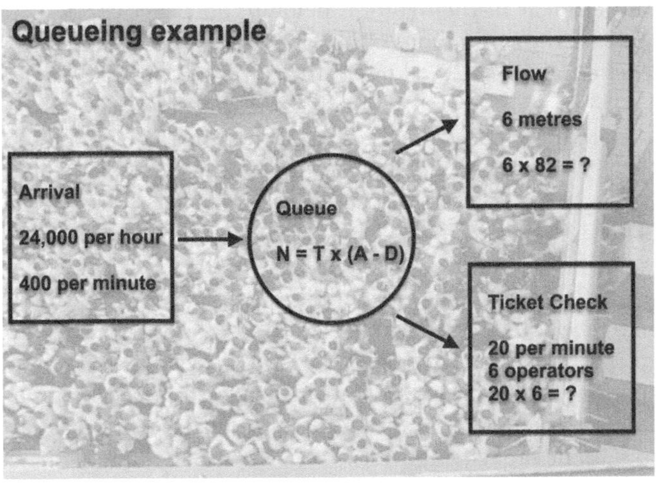

Figure 8.4 Combined flow and ticket check.

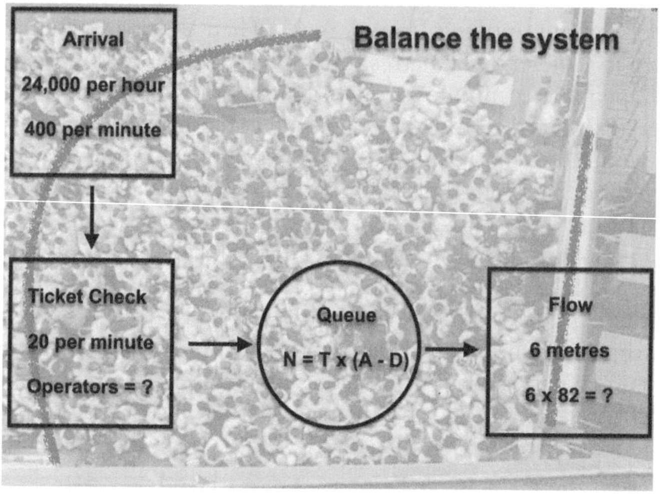

Figure 8.5 Separate ticket check and flow.

There is a ticket checking process at the narrowest point with a service rate of 20 tickets per minute (times 6=120 patrons per minute). If there were no ticket checks, then the flow rate would be 492 patrons per minute through the narrowed point. The ticket checking process is the slowest element of the system and the arrival rate at 400 patrons per minute exceeds the ticket check rate of 120 patrons per minute. Each minute the queue will grow by 280 patrons.

If we separate the ticket checking process, move it further away from the narrowest point, we can make this a more efficient system. Creating a line (the curved element in the diagram below) and ticket check at that point, we can increase the throughput of this system. We need to increase the number of ticket checkers to 20, along the curved line, to equal the arrival rate, then the patrons will flow through the gates. Of course, you may need to add 20% more ticket checkers, for contingency, in this example (Figure 8.5).

SUMMARY

The use of a simple queueing formula is a first pass approximation. It will allow you to check if a system can function. If it fails a first pass approximation, then the system requires redesign. If your system works in a first pass approximation, then you can look at the process in more detail. The objective is to determine works or does not work, with the first pass approximation or rough-cut capacity analysis.

Chapter 9

Crowd risk analysis

> Risk comes from not knowing what you're doing.
>
> Warren Buffett

> What a strange sense of logic to fixate on the absence of something (injury) as a demonstration of the presence of something else (safety). Such a proposition misunderstands the dynamic of risk and being human.
>
> Dr Robert Long

In this chapter, we outline both the public and professional perceptions of risk. Risk reporting needs to describe risks that change over time. We demonstrate a method, used for over 20 years, to outline the location, duration and severity of crowd risks.

WHAT IS MEANT BY A "HAZARD", A "RISK" AND A "THREAT"?

First, we need definitions, of risks, hazards and threats. From the Safety at Sports Grounds Authority (UK) – SG03[1] Section 2.2

a. A **hazard** is something that has the potential to cause harm to people. A hazard could be a physical element, such as a damaged fence or a poorly designed stairway. It could be a void or a slope, or it could be a fitting, such as a faulty light, a banner or a sign. A hazard might also be a material, such as unprotected glass, or a substance, such as liquefied petroleum gas (LPG). Hazards can also arise from the activity taking place at the venue, such as balls striking spectators, or pyrotechnics

[1] https://sgsa.org.uk/draft-supplementary-guidance-03-event-safety-management/sg03-event-safety-management-draft/

DOI: 10.1201/9781351053068-9

let off as part of displays. A further hazard is of course an infectious virus, such as COVID-19.

b. A **risk** is the likelihood that a hazard will cause actual harm. Some hazards may cause harm, some may not. It is essential therefore for the venue management and/or the event organisers to:
    i. identify hazards
    ii. assess their potential to cause harm, and
    iii. take the steps necessary either to remove the hazard or to mitigate the potential impact of that hazard.

        These measures are generally referred to as "risk assessment" and "risk management". Indeed, it could be said that all safety management is, in essence, about the assessment and management of risk.

c. A threat is generally understood to be a statement expressing an intention to cause harm; be it physical harm to a structure or person, financial harm, reputational harm or cyber harm. A threat can also be a person behaving in a threatening or suspicious manner, either in person at the venue or its environs, or online by sharing threatening material.

A TYPICAL RISK ASSESSMENT

For crowd safety, this would form the risk assessment, a list of hazards, usually in no specific order, coupled to a list of risk mitigation/management options/choices. Table 9.1 is an example from the UK HSE, Health and Safety Executive[2].

The risk assessor would complete this list, typically in random order, trying to think of the possible hazards and risk mitigation. The problem is that this requires an effort to complete. You need to write/type the risk descriptions and then consider the various risk mitigation options. We often see risk assessments copied from event to event and filled in with little thought to the implications in our expert witness cases. In some organizations, this can become a box-ticking exercise[3].

You should note that the risk assessment is one of the first documents an expert would scrutinise following an incident. The expert would check to see if a specific risk was overlooked, underestimated, or ignored. If the documentation failed to recognise the hazard, failed to take appropriate risk mitigation, and/or did not adopt best practices to manage and risks/hazards during the event.

---

[2] https://www.hse.gov.uk/simple-health-safety/risk/risk-assessment-template-and-examples.htm

[3] https://www.bbc.co.uk/news/uk-england-manchester-54809878

Table 9.1 HSE sample risk assessment

What are the hazards?	*Who might be harmed and how?*	*What are you already doing?*	*Do you need to do anything else to control this risk?*	*Action by who?*	*Action by when?*	*Done*
Slips and trips	Staff and visitors may be injured if they trip over objects or slip-on spillages.	General good housekeeping is carried out. All areas well lit, including stairs. No trailing leads or cables. Staff keep work areas clear, e.g., no boxes left in walkways, deliveries stored immediately.	Better housekeeping in staff kitchen needed, e.g., on spills. Arrange for loose carpet tile on second floor to be repaired/ replaced.	All staff, supervisor to monitor Manager	—	—
Next hazard	—	—	—	—	—	—

The problem is this process is both prescriptive, i.e., this is a legal requirement in the UK, and restrictive, in that it is difficult to put enough words in the template to define the specific risk. As the HSE website[4] states,

> Do not just copy an example and put your company name to it as that would not satisfy the law and would not protect your employees.

We would also add *do not underestimate the amount of work/time required to complete a proper risk assessment.*

PUBLIC AND PROFESSIONAL RISKS

An UK Utility company engaged us to review their roadworks accident record, to review why people fall into holes in the road? During the national accident statistics review, they did not appear to have a higher number of incidents than other similar companies. However, there was a perception in the media that they had a poor record for public safety.

We had the companies' risk assessments, accident and incident reports, and site photographs of each location. The photographs provided an important clue to addressing the issue. Messy roadworks had fewer incidents than tidy roadworks. This was counterintuitive; why would more ordered sites have a higher frequency/number of accidents?

We called this the perceived risk project.

UNDERSTANDING RISK

To understand risks to the public, you need to see the hazard from the public's perspective. A messy, untidy site, looks dangerous, looks like a high-risk environment, hence individuals take care navigating the area. Whereas a tidy, ordered site has the *illusion* of negligible risk. The risk of tripping and falling into a utility works has the *same* risk, messy or tidy, but the public's perception of risk is going to be different. When we voluntarily take a risk, we usually *underestimate* the chances of hurting ourselves. To *understand* risk, we must understand the *probability* of danger *and* the *consequences*. This is how the professional understand risk, but the public may simply think "Hold my beer, watch what I can do" and fail to understand the probability of danger and consequences of their actions.

A further revelation during the project was the accident reporting, which usually started with "MoP[5] fell into a hole on the road" without evaluating if there was an underlying reason for the accident.

[4] https://www.hse.gov.uk/simple-health-safety/risk/risk-assessment-template-and-examples.htm

[5] Member of the public.

THE MAGICIAN'S EYE

Following our initial research, during the first project meeting, we presented a series of coin vanishes and transpositions[6] to demonstrate both sleight of hand and sleight of mind. If you watch the video, the ring does not *actually* change into the gold coin and back (several times), but the illusion is that *it does* change. It was the study and practice of creating such illusions, knowing the audience's perspectives, understanding, and using the lines of sight knowing how colours and objects can complete the illusion, which gave us insights to the client problem.

When we researched the client's accident logs, and accident site photographs, we found lighting, shadows, colours, positioning, tidiness/order on a site. All contributed to why there was no perception of risk. This was key to improving risk assessment on their site and their accident reporting logs. This is a quote from the client's:

> I was privileged to work with Keith on a 'Perceived Risk' project with regards to injuries to operatives and members of the public for (the Utility company). The input from Keith regarding people's general perception to risk was invaluable. His knowledge is extensive and in depth and highlighted important factors that may normally go unidentified. His unique approach, background and theories made for interesting research for me and led to some 'out of the box' proposals being put forward. Keith was a great inspiration to me as a developing engineer and I have taken some aspects of thinking with me into my following roles.
>
> Dave Vallender.
> *Civil Engineer.*

CROWD RISK ANALYSIS

Risk analysis is one of our favourite topics in our courses, mostly because it highlights the difference between the public perceptions of risk and the professional approach to risk assessment. We have a wide range of photographs and images of people doing the stupidest of high-risk activities, oblivious of the risk they are exposing. The Darwin Awards[7] are full of examples. People have been sending us examples, from around the world, for years, and it never fails to provide laughter in the classroom, and few sheepish grins from those who recognise a bit of themselves in some of the photographs.

---

[6] https://www.gkstill.com/CV/Magic/Spellbound.html
[7] https://darwinawards.com

RISK PERCEPTIONS 1 – ROULETTE

You need to understand that public's perception of risk can be incorrect. An example of this is the vaccination opposition (or anti-vaxxers)[8] or flat earthers, who, despite the evidence, will not shift from their belief systems[9].

We also need to understand that, as professionals, how we assess risk. Finally, we need to balance our understanding of risk with the public's understanding of risk. In other words, your risk assessment should consider both the risks and the public's perception of risk. To illustrate this, consider this scenario.

> You are playing the game of roulette and, after nine consecutive reds in a row (the odds for that are 600 to one), do you bet black?

For the purposes of this example, we can assume the roulette wheel is fair, and that there is an even chance of red of black on the next spin. So, do you bet black? You might think, *yeah, black has got to come up next* however, the odds for the next spin have not changed, the wheel has no memory, it does not know about nine reds in a row, the odds are still 50/50 red or black. The illusion, the sleight of mind, tricks you into thinking that you have higher chance of winning betting black, because you think that nine reds in a row is relevant. The same principle applies to National Lottery "predictions" based on low frequency of numbers. We tend to see patterns where patterns do not exist. That is why the casino always wins[10].

RISK PERCEPTIONS 2 – THE BRIDGE

Consider the bridge in Figure 9.1, would you cross this river, using this bridge? It looks dangerous.

If we offer a prize of one million (dollars or pounds) to cross this river, using this bridge. How do you now perceive the risk? Obviously, the risk has not changed, but the reward can shift your perceptions, you might think, it is worth taking that risk, for the reward.

[8] https://www.healthline.com/health/vaccinations/opposition

[9] https://www.who.int/news-room/q-a-detail/vaccines-and-immunization-myths-and-misconceptions

[10] https://www.bbc.com/travel/article/20120816-travelwise-casino-design-and-why-the-house-always-wins

Figure 9.1 Would you cross that bridge.

RISK VS REWARD

In these two examples, the roulette wheel and the bridge, we can see that risk is not simply a calculation of probability; it is also a perception. To understand crowd risks, you need to understand this from a shift in *your* perception, and you need to understand risk as seen from the *crowd's* perspective.

IF IT IS PREDICTABLE, IT IS PREVENTABLE

As an expert witness, we look at the criteria of the foreseeable. Is there a causal chain of events that could lead to an accident or incident? For example, in an accident investigation, we can ask; was the risk overlooked, ignored, underestimated, known, and accepted or were there risky processes and procedures? We can also ask, did the event organiser, or employer, understand that stupid people can do stupid things? Anticipating crowd behaviour and mitigate those risks is an essential element of the risk assessment.

A PICTURE PAINTS A THOUSAND WORDS

For a crowd risk assessment, you need to understand the motivations of the crowd. For example, if you design a queuing system or method to regulate the crowd flow into space, you need to understand the impact of waiting, specifically close to the start of an event, and how that may affect crowd behaviour.

We also need to consider the dynamics of risk. Risk can be variable, for example, overcrowding at an entry point. Two hours before the event starts, there may be zero crowd risks (no-one on site). One hour the risk may be low, as crowds begin to enter the site. Fifteen minutes before the start of the event, where queues have accumulated and people may be anxious about getting in, to their seat/position, that same area may see a higher risk for overcrowding and potential crushing.

We need to consider the risk locations and how long that risk might exist at that location (duration). We need to understand the severity of the risk, how that may change over time. A specific location might be a risk of overcrowding which could be high or low at various times. All of this would require documenting in your risk assessment. The nature of the risk, the location of the risk, the duration of the risk and the severity of the risk. Consider writing this in words, consider how to describe the location, duration and severity of various crowd risks in and around your site, in your risk assessment, then consider if there is a method of capturing the same information in a picture.

VISUALISING RISK

Think about the process of risk assessment. You first imagine/visualise the risks. Then you need to articulate and document the risk. Then someone else picks up the risk assessment and tries to visualise the risk of implementing the risk mitigation. There are various stages involved in this process, all prone to information loss or misinterpretation. There is a simpler method to describe the risk, pictures (Figure 9.2).

These are risk/congestion maps, using the Red/Orange/Green to show 1–5 people per square metre/yard (Chapter 2 – Crowd Dynamics). If you add a timeline across the top (Figure 9.3) you have, in a single diagram showing risk/congestion and time. Several diagrams like this will produce a timeline of risk due to congestion.

Using diagrams, we can visualise the risk duration, showing changes in severity, at locations, over time. When we couple this with the DIM-ICE (Design, Information, Management, Ingress, Circulaiton, Egress) risk models, we capture the dynamics of crowd risks.

THE FUTURE OF CROWD RISK ASSESSMENT

In countries/states where there are statutory requirements to produce a risk assessment, the standard forms have been in circulation, unchanged, for decades. We are not advocating dropping or replacing those forms. The conventional risk assessment process, engrained in both the culture and the process/procedures, for the local authorities and organisations should

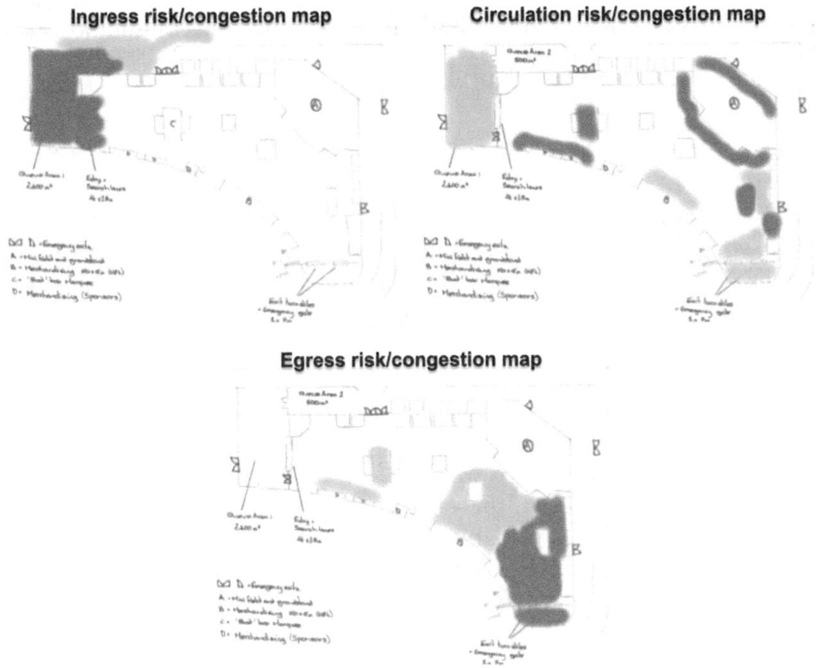

Figure 9.2 Three sketched diagrams (ICE) showing congestion.

1000	1015	1030	1045	1100	1115	1130	1145	1200	1215	1230	1245	1300	1315	1330	1345	1400	1415	1430	1445	1500	1515	1530	1545	1600	1615	1630	1645	1700	1715	1730	1745	1800	1815	1830

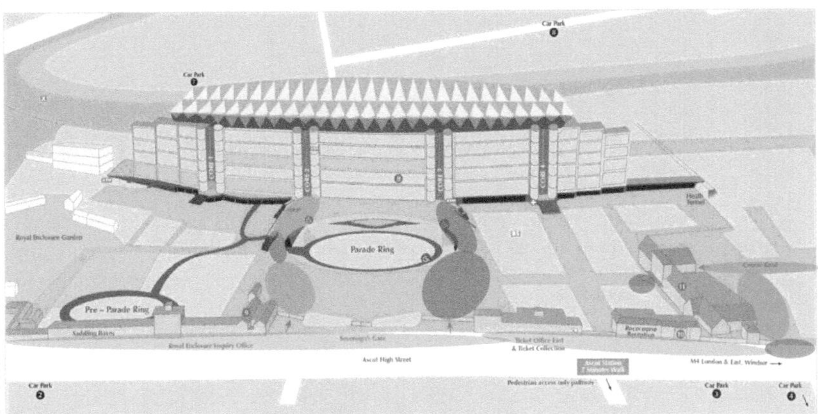

Figure 9.3 Dubai Mall – time and congestion map.

not change overnight. We *are* recommending that you should use supplementary images/graphics, to aid both understanding and communication of your crowd risk assessment.

VISUAL TO VISUAL

Crowd risk assessment starts and end with a visual process. As previously stated, you begin by trying to visualise the crowd risks. In a few words, and a single number, you document those risks. Later, someone else must pick up that document and then try and re-visualise the risk. You can produce a series of graphics that are easier to create, document, visualise, understand, and act upon the information contained in the graphics. Adding that to your statutory risk assessment is simple, cost-effective, informative and can provide more information across multi-agency teams for managing places of public assembly. We called the process risk and congestion mapping. As you will see in the various case studies, coupling this with a RAMP analysis provides valuable insights to the location, time, direction, and risks associated with the crowd dynamics on large sites. understanding and mitigating risks is the key to crowd safety.

Chapter 10

Event egress analysis

> Hindsight is a wonderful thing, but foresight is better, especially when it comes to saving life, or some pain!
>
> William Blake

In the previous chapters, we outlined a set of tools for assessing crowd risks. In this chapter, we will put these together to analyse event egress analysis. To understand your limits of liability, you will need to understand your boundaries. You need to consider both your legal responsibility for the crowd departing your site, and the potential reputational damage that may result if there is an incident during egress, and your event/brand is associated with that incident.

Typically, these details are in you site licence/permit, defining the terms and conditions of the site usage and the potential risks. The principles we shall demonstrate in this section apply to the end of event egress and also to any site migration, from one area to another, from one stage to another, mid-event.

EVENT EGRESS

We were a prosecution expert witness in the Falls Festival incident (2016) and the subsequent successful, multimillion dollar class action lawsuit.[1,2] In the article, "Falls Festival pays out first stampede victims in multimillion-dollar class action" there are two diagrams[3] which highlights the incident. The diagram shows a narrow point in the system. This is typical of the problems we see in system design, where ingress over a longer period is low flow, but egress occurs in a much shorter time, for the same number

1 https://www.supremecourt.vic.gov.au/court-decisions/case-list/falls-music-and-arts-festival-class-action

2 https://www.gadens.com/legal-insights/falls-festival-class-action-claimants-lawyers-win-contingency-fees/

3 https://www.theage.com.au/national/victoria/falls-festival-pays-out-first-stampede-victims-in-multi-million-dollar-class-action-20190611-p51wmy.html

DOI: 10.1201/9781351053068-10

of patrons. Therefore, a much higher density/risk is going to occur during egress. It will be the same at the end of an event, people arrive over an extended period, but leave over a much-abbreviated period of time.

However, if you know the maximum capacity (flow rate/density) of any point in the system, and the number of patrons that may move through that point, then you can calculate the time it will take. Also, if there is limited time, if the number of people (the demand), exceeded the capacity (the width of the doors/gates), then you can anticipate problems and potential injuries. Using the calculations, we can then work out what flow/volume/capacity we need to provide for safe passage, for any part of the system.

FORESEEABLE

Why this is important is that, in legal cases, the criteria of "was the accident/incident reasonably foreseeable", hence, was it predictable, was it preventable, are the criteria we apply to incident investigation. So, an expert would apply the type of risk assessment from this chapter, and present that as evidence of reasonable foreseeability, will demonstrate that the incident was predictable, and hence preventable. It may not be necessary for an expert to demonstrate how an incident could be prevented, merely that it was predictable.

If there is a foreseeable potential for a crowd crush, then the incident is preventable, then the risk assessment should have identified the areas of concern. Armed with that knowledge, it is up to the event organiser, to make safe those areas. There are, as we have highlighted throughout this book, options to make spaces safer through design changes, changes in scheduling or information, or changes in the management of the crowds. The principles are always to first identify the problem, identify the nature of the problem by applying the risk management principles, then (and only then) can you apply your team to working out a solution.

CLASSROOM EXERCISE

In the courses, we use a specific site, one that had an issue, and demonstrate how modelling helps identify, then resolve the problem. We use a site from 1999, Morpha Stadium[4] Stereophonics concert as a site egress example. The event was a success[5] for the band, but there were challenges for crowd safety during egress. Fortunately, there were no serious injuries. The safety officer took the site plans to one of our workshops and we have used it

[4] https://www.youtube.com/watch?v=9qGxXT9Jm60

[5] https://www.walesonline.co.uk/whats-on/whats-on-news/stereophonics-iconic-gig-morfa-stadium-15534311

in courses as an example of how to apply egress crowd safety analysis. Figure 10.1 shows the site.

There was a railway track on one side, a river on the other and our event space, is the oval running track (middle, bottom, left image, middle top, right image). The calculated site capacity is 50,000 people. When they leave the event, they are going to go to use the route shown below, following the road, towards the bridge and the shuttle bus area. The area is closed to traffic, except for a red route (emergency access) which uses the same roads.

Due to the site location, river along one side, railway line along another and a wooded area along the other side, the crowd only really have that one route on and off the site.

Figure 10.1 Two images of the site.

The site had a total open width space of 40m for egress, which consisted of a series of gates. With these dimensions, we can apply a calculation. 50,000 spectators at 82 people per meter per minute on a 40-metre-wide gate system should give us, 3,280 people per minute flowing through the exit. As a first pass approximation, it would take 15 minutes and 15 seconds to get all 50,000 through that 40-metre-wide gate system. Typically, that is as far as they applied their site egress calculations; the boundary of the site. However, if we then look at the section beyond the gate system, we can see a problem. The roads (Figure 10.1) look narrower than the 40-metre-wide gate system. The road that runs between the Arena and Car Park 1.

BOTTLENECK

However, the problem here is that the width of the roads that go north and south is 12 meters wide. So, 40 metres of crowd flow will require 40 metres of egress width, along the route. There is a maximum of 24 metres (assuming 50–50 split of the crowd going North/South). There is a section of road, 100 metres in length, before the South bound 12-metre-wide road, turns along the emergency vehicle access route (dotted line above). At that junction the road reduces to 6 metres wide as the road was divided along its length, using drag barrier, to separate the traffic and the crowd. As we can now see, the exit gate (40 metres) is not the constraint in the system.

This can be difficult to understand from a written description. It is also difficult to visualise if you are looking at aerial site images, as they all show the site from different perspectives. Modelling will help us understand and communicate the areas of concern for this site.

MODELLING THE SITE

We need to break the problem down to its simplest geometry, and then apply a first pass approximation to understand the potential problems.

An estimated 50,000 will attend the event. The Exit gate is 40 metres wide (edge of the site boundary) and the road is 12 metres wide, to the North, and 12 metres wide to the South, with a 6 metres wide section after 100 metres. From the ticket sales, we know that the majority (estimated 80%) will approach the site from the South. 20% will be from the local area, and it was expected that they would use the Northern approach road.

To model egress, we need to consider the areas around the system constraints, the 40-metre-wide egress, the 12-metre-wide road and the 100-metre section to the 6-metre-wide junction. Figure 12.1 shows the area

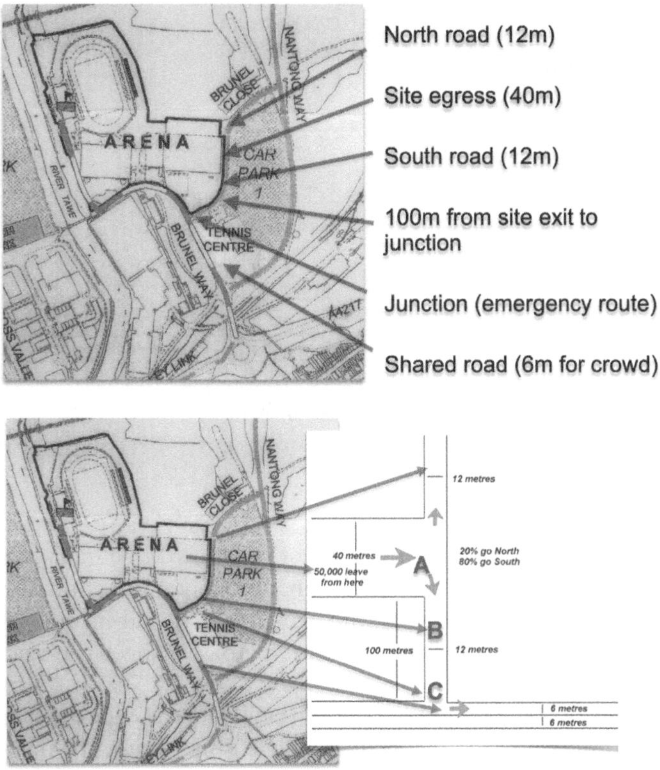

Figure 10.2 Key dimensions and line model of the egress system.

we need to understand, for density, flow, and time for egress and how we can convert that to a simple egress line diagram. When creating these models, you simply ignore everything that is not directly relevant to your analysis (Figure 10.2).

THE SITE EGRESS MODEL

In this kind of problem, we need to assess what will happen when a crowd, flowing out of the 40-metre-wide event site, turns North (20%) and South (80%) then, after 100 metres the road narrows again to 6 metres. We can make some approximations in that 20% of the 40 metres of crowd flowing out of the gates will turn North. That is an estimated 8 metres (20% of 40 metres). The road is 12 metres wide; therefore, we do not anticipate a problem for the crowd heading North as the road is wider than the anticipated flow and percentage distribution.

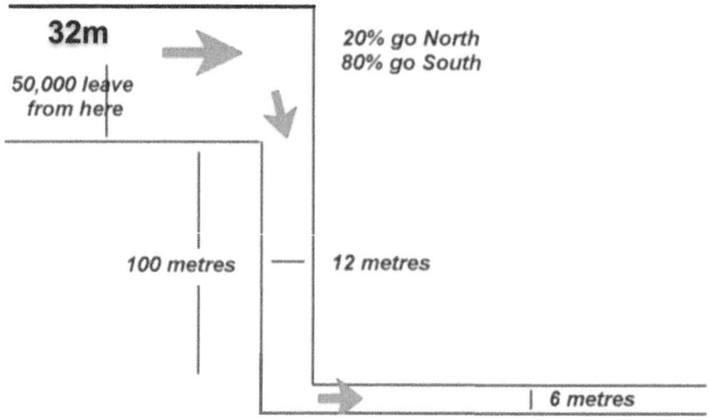

Figure 10.3 Simplified line model of the egress system.

SIMPLIFICATION

Of course, this is an oversimplification of the crowd dynamics, but that is the nature of a first pass approximation. We make an estimate based on simplified geometry assumptions, the principle is that if it fails a first pass approximation, there is a system problem, if it passes this basic test, then we will refine the model further.

So, 8 metres of the crowd flow, out of the 40-metre-wide gates, will flow North, leaving 32 metres (40–8 metres) flowing South. That is an estimated 32 metres of crowd flow, trying to move onto a 12-metre-wide road. It is not going to work; QED. However, the problem needs another step. After 100 metres (100 metres times 12 metres wide is 1,200 square metres) the road narrows again to 6 metre wide. The crowd that flows down the 12-metre-wide section now reaches a further constraint. The road narrows again. We can simplify our diagram, ignoring the North and show the site without the North bound crowd flow (Figure 10.3).

FURTHER SIMPLIFICATION

Then we can simplify that diagram, by straightening out the geometry: 32 metres, down to 12 metres, then down to 6 metres. Typically, this kind of simplified diagram should be enough to cause the event organiser to rethink their site design (Figure 10.4).

It may be necessary to show the scale of the problem, so we can refine the model and add some calculations. Then we can create a spreadsheet (Figure 10.5) of the above, showing minute, by minute, how the crowd congestion will build up.

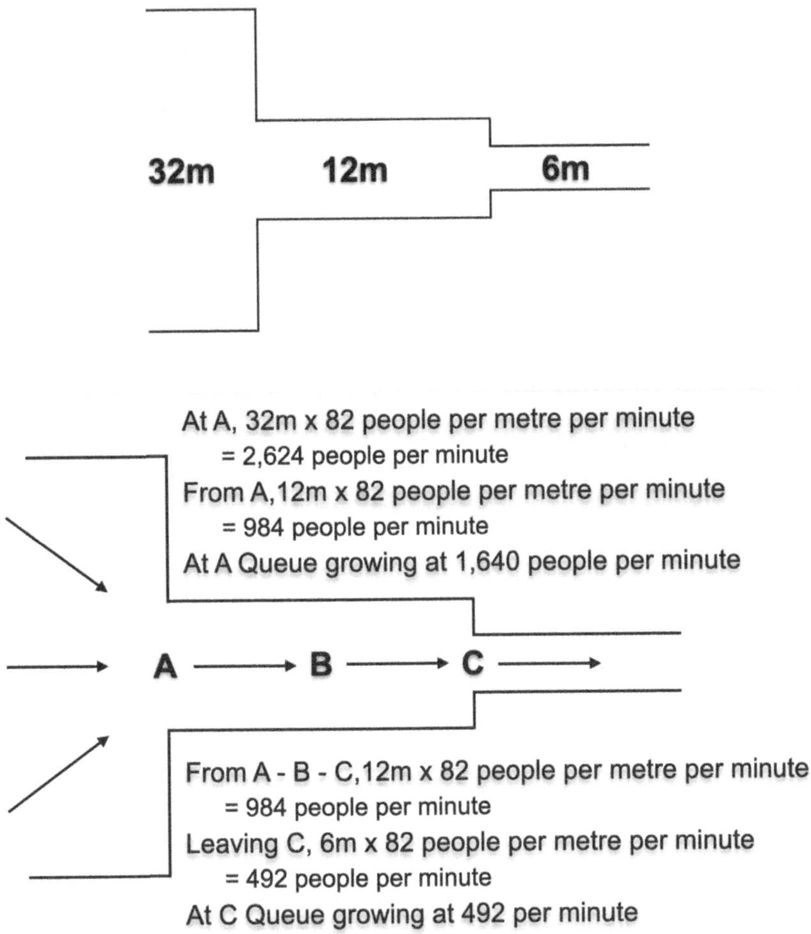

Figure 10.4 Simplified line diagram and sample flow calculations.

SPREADSHEET ANALYSIS

When we produce this type of spreadsheet, with stated assumptions down the left-hand side, and time across the top. We can create a graph of the crowd build up. Now, we can see the time it takes for the 1,200 square metres section to reach four people per square metre (just over 9 minutes) and the total number of people in the system and see how the crowd risks will develop over time.

Many of the worlds' crowd disasters could have been avoided had they used this approach, before their event.

Initial number of people in the park	50000									
Time progression (in minutes)	1	2	3	4	5	6	7	8	9	10
Passing point A (40 metres) not capped (82*40)	3280	3280	3280	3280	3280	3280	3280	3280	3280	3280
People still in the park	46720	43440	40160	36880	33600	30320	27040	23760	20480	17200
People arriving at point B (80%)	2624	2624	2624	2624	2624	2624	2624	2624	2624	2624
People passing point B	984	984	984	984	984	984	984	984	984	984
People arriving at B and queueing at A	1640	3280	4920	6560	8200	9840	11480	13120	14760	16400
People in the park (due to B)	47376	44752	42128	39504	36880	34256	31632	29008	26384	23760
Number of people that can pass point C	492	492	492	492	492	492	492	492	492	492
People arriving at C from B	984	984	984	984	984	984	984	984	984	984
Queueing at point C	492	984	1476	1968	2460	2952	3444	3936	4428	4920
People in the system	49508	49016	48524	48032	47540	47048	46556	46064	45572	45080
Capacity at B = 100 * 12 metres (1200)										4800

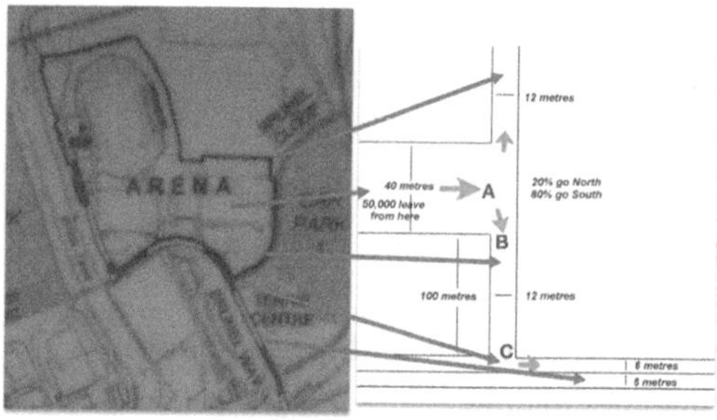

Figure 10.5 Spreadsheet numerical analysis.

EGRESS DYNAMICS

Initially, the 40-metre-wide gates will not be flowing at full capacity. The crowd will initially move North and South, at lower flow rates, as some people leave early (to avoid the traffic rush). After a few minutes, the congestion will start to build up at section C, then back up to the Gates and then the entire system will develop into a crush zone. The crowd dynamics for this area is reasonably foreseeable, predictable and hence this crush zone is preventable.

Simplifying the site design helps the user visualise the problems. When we run this in the classroom, we stop at the site, ask the learners to develop a site diagram by themselves. This is the most essential element of modelling, simplifying the site. What this achieves is to reduce the site to just the key elements of the system. If it fails a first pass approximation, the site needs a redesigned. If the site looks like it might work with a first pass approximation, you can test it further, using the queuing formulas, or I can use risk and congestion mapping. There are people who prefer the written descriptions, some the numerical/spreadsheet version, some the risk and congestion mapping, some prefer to see the animation of the congestion building up at the 40-metre-wide gate, along the road to the 6-metre-wide constraint. We are not all the same, that is why we use several types of models, to help communicate risks, and help the event organiser, or local police, see the problem for themselves. If you present the problem using different visualisations, whether it is a colour diagram, the calculations, a spreadsheet, a block/line diagram, these are all different techniques that allow us to visualise and evaluate the risk.

REDESIGN THE SITE

In the classroom, we then task the learners to redesign the site. They use the models and present their solution to the event organiser. There is a sleight mind technique you can use for budget approval. You may have costed your solution at (say) £6,000. If you present only that option in your budget proposal, you can expect a "can you do it for less?" If you present two solutions, one that is expensive (in this case, that might be to build a bridge) and the other that you have worked out (your £6,000 solution) then the choice is A or B. It will be easier to sell the £6,000 option. Always present an "expensive" option first.

If you want to see those solutions, the course is available online, and we walk you through each stage of the problem, with animated diagrams, and through the solution.

Chapter 11

Strategic and tactical analysis

> Nothing is more central to an organisation's effectiveness than its ability to transmit accurate, relevant and understandable information among its members.
>
> Professor T.J. Larkin

In this chapter, we outline the development and application of games theory models for crowd safety and risk analysis. We illustrate its use for planning and analysis of crowd safety, crowd risk analysis and both strategic and tactical options for control room references and for decision support during emergency situations.

In the previous chapters, we outlined the DIM-ICE risk model, the principles of crowd dynamics (static, dynamic and contingency spaces), we covered the applications of a RAMP analysis and outlined the use of risk/congestion mapping tools. Those are four of the key tools we developed to assist in the process of crowd risk analysis and crowd safety engineering. The fifth of our tools is the Strategic and Tactical analysis tool (Decision Support Analysis). This is derivation from Games Theory model.

BACKGROUND

Game theory[1] is "the study of mathematical models of conflict and cooperation between intelligent rational decision-makers". John von Neumann originally developed the concept from the idea that certain strategies players adopt can lead to successful outcomes. Game theory is the science of logical decision making in humans, animals and computers.

[1] *Game Theory: Analysis of Conflict*, Harvard University Press, 1991.

DOI: 10.1201/9781351053068-11

CROWD SCIENCE AND GAME THEORY

The process that organisations typically adopt for decision-making is a list approach. For example, under scenario A, we have the options X, Y, Z. For scenario B, we have options W, X, Z. This type of approach can be highly subjective, selective and can be restricted in its application.

Before we outline the concepts of using Games Theory for strategy and tactics, we need to explore why this is an important aspect of crowd science. When first applied these principles, in the early 1990s during his research at Wembley Stadium, there were conflicting signs, crowd flowing in odd directions and, in general, the crowds did not understand the information. The question was "why don't people follow the signs?"

The answer is that the signage is information, it influences the crowd dynamic. If your information fails to connect with both the eye and the expectations of the crowd then it serves no purpose. An example of this, Figure 12.1, it is perfectly clear to the staff what those arrows mean, but not to someone who sees this for the first time. We have hundreds of similar examples, from venues around the world. Signage is a communications system; it is how the site/venue/event informs the crowd (Figure 11.1).

COMMUNICATION FAILURE

Signage is a method to communicate with people, with the crowd. When we communicate expecting an action or reaction, there can often be a disconnect between the sender and the receiver. When we communicate with a crowd, this can result in mass action or reaction, therefore it is

Figure 11.1 Stadium signage.

important to consider the process of crowd communication. As William White once said[2], "the greatest barrier to communication is the illusion of it in the mind of the sender". The quote is often, incorrectly, attributed to George Bernard Shaw as "The problem with information is the illusion that it is complete". To understand how to communicate with a crowd, you need to understand the process of communication, and the diverse ways that communication can fail.

PHENOMENOLOGY

We perceive the world through our senses. We can easily fool our senses fooled; ask any magician, but do not ask a conman, or you *will* lose your wallet. The principle of any illusion follows a simple philosophy, in that we can shape your perceptions by what you see, what you think you see, what you perceived about what you thought you saw (interpret)ed, what you think you remember about what you saw. Then amplified in the stories you tell others about what you thought you saw. It is that chain of thought that that magician manipulates to create an illusion. Derren Brown in his book, *Pure Effect*, tells of a conversation where someone told him of a trick, they had seen 20 years before, in a bar. A chord struck with Derren, he wanted to create a talked about illusion, 20 years from now.

In "Mind Control" season 2, episode 1, Derren turns the table on advertising executives, by predicting how they will shape their campaign, draw their advertising image and slogan, by embedding information into the environment, using subliminal advertising. Those same techniques the advertising executives use on members of the public to create a sense of familiarly for a product they think they have never heard or seen before. Millions are spent on marketing campaigns, and we all fall for it. Browse any bookshop, or search on the internet. You will find books on marketing, contain chapters on crowd effects, use a wide range of associations and play with your senses and memory. We can use the same techniques for embedding safety information into the environment. First, we need to understand how the mind perceives and records data.

VISUAL PERCEPTIONS

When lecturing, we often illustrate our talks with illusions, to demonstrate how easy it is for a performer to shape the audiences' perceptions. We use visual, tactile, aural and subliminal components within the course materials, to emphasis key safety elements.

2 1950 September, Fortune, "Is Anybody Listening?" by William Hollingsworth Whyte, Start Page 77, Quote Page 174, Published by Time, Inc., New York. (Verified on microfilm).

One example, Ninio's extinction illusion[3] where the viewer cannot see all 12 black dots at once leads us into a discussion on CCTV camera operators, what they can see, what they perceive. This is another example of how the magician performs, understanding their art, using the small area your eye can "actually" see in detail. We often illustrate this in class with a pack of cards. Sit a volunteer down on a chair, facing the rest of the class and stand behind them. Randomly select a card from the pack. Tell the volunteer to face the audience and not to look round (assure them that nothing is going to happen). Slowly, move the card from the far left/right round the volunteer's head. Ask them to identify the card.

You will note that the volunteer can detect movement at around 90 degrees (left/right) but not colour or form. Move the card, in a circle (about 50 cm/2 feet out) and ask if they can see any colour, or shape. If the volunteer does not turn their eye towards the card (you can evaluate this on yourself) they will be able to tell colour at approx. 45 degrees and the face of the card at around 5 degrees of arc. This surprises most volunteers, specifically if you ask them about how good their vision is before the experiment. Of course, this is the narrow perspective we exploit in most close-up, parlour, and stage magic. Knowing the angles, the field of view and the persistence of vison are all essential to create a perfect illusion.

This, along with other illusions we highlight how easily our senses can be deceived. These demonstrations also serve to illustrate why the use of crowd models can help heighten our awareness of crowd risks, both from the control room perspective and from the ground level view of the crowd. Lines of sight, colours, language, the use of words, all of these are important to understand when we plan wayfinding towards, around and away from a site/venue/event.

Signage is the principal method of directing the crowd around complex spaces, but it if not noticed, if it not seen, if not understood and not acted upon, it fails to serve its purpose, it fails to influence crowd behaviour.

SHAPING CROWD BEHAVIOUR

Our work has focussed on shaping the crowds' behaviour and expectations, on directing attention through the design of the site, via advanced notifications, via social media and via other information channels. You can affect crowd behaviour through your management processes and procedures, and colour of uniform. For example, the black clad, heavy set "security" guard projects an image in the mind of the people attending an event. An experiment, at Bucks New University, using pastel shades and replacing the word "security" with the word "safety" on

[3] Ninio, J. and Stevens, K. A. (2000) Variations on the Hermann grid: an extinction illusion. *Perception*, 29, 1209–1217.

the back of shirts, elicited a greater compliance, and approachability, between the staff and the crowd. Simple concept, but highly effective to aid communication.

WHAT IS A COMMUNICATION FAILURE?

The key to understanding the principles of crowd science as it applies to event safety is in understanding how the crowd receives and perceives information. As we have outlined in previous chapter, if we are managing the audiences' expectations, if we are shaping their perceptions, how they receive and process information, then we can apply those same principles used to create stage illusions, to event safety. To understand why this is important, we need to outline communication failures. This has a hierarchy.

First, it could be that you do not provide any information. Without instructions, people may not know what is expected of them in any given environment. An example of this could be a series of directional signage toward a site/event/venue. You begin to follow the signs, and at a critical junction – a sign is missing. If you cannot determine which junction to take, there may be an immediate frustration, particularly if time is running short before an event starts. Lack of information can lead to frustration, and this can shape a crowd's behaviour, not just at that instant, but it can create an atmosphere, a mood, an annoyance, which persists. Ensuring approach information is clear, consistent and visible can negate that effect. Also, providing pre-event information, how to get to the event, how long it might take and which mode of transport consider how your site information may help reduce ingress frustrations.

At the next level, you could provide/transmit information, but it is not received. A sign, not seen or an announcement not noticed would be examples of information given, but not received. Then we might have information given, received but not understood, for example, a public announcement in a strong local dialect. Information can be given, received, understood, but not remembered. Finally, Information may be given received, understood, remembered but not acted upon. An effective communication system needs to consider information these elements of potential failure, in easy to understand, easy to remember and about all clear packets of information. Words matter, language is important, and we covered the use of words, the use of colour, of diagrams, of line drawings and flow diagrams, in previous chapters.

JARGON

> The chief virtue that language can have is clearness, and nothing detracts from it so much as the use of unfamiliar words
>
> Hippocrates

Our ability to communicate with the crowd depends on the use and understanding of the language and nothing detracts more than the use of unfamiliar words. The various agencies that operating around an event site use jargon, for example, silver control, orange car park, loading point, rendezvous point, pickup point. Even left or right depends on which direction you are facing at the time. Main entrance, inner concourse, main stage and terms that are familiar to the operators on a site, but not necessarily familiar to the crowd. This can detract from the message that you are trying to communicate. An example, where this becomes critical, is during an evacuation announcement.

> Can everyone please leave by the Wessex Gate

Is effective if, and only if, you know where the Wessex Gate is located. Information given, received, understood but not acted upon because it relies on the crowd knowing the location of the Wessex gate. In other words, the information is incomplete, you have failed to communicate with the crowd. From our experience:

> My son, Harry, was eight when he came with me on a site survey for Lincoln Christmas Market. There was a public announcement that stated, "If you become separated from your party, please assemble at front of the cathedral." Harry, asked me, "What does it mean? Where is the party and is there going to be cake?" I tried to explain that it meant a group of people. He said, "So why didn't he say that?" it was a good point. So, we went up to the control room and asked about the message? The announcer told us that it was for lost kids and Harry, bless his heart, said to him, "Why aren't you using a language that kids can understand? Like, kids, if you're lost, see a bloke in a yellow coat[4], they'll help you out"

It is essential to understand that when you are communicating, who is receiving the information, what actions do you require, will they understand the message, will they remember the message and will the act upon the message. If individuals, and hence the crowd, do not see, understand, and act upon this information, then you have failed to communicate. This may seem trivial, but in an emergency, that information, such as exit information and site orientation, can save lives.

HOW DO WE ASSESS OUR COMMUNICATIONS?

How do we analyse our communications systems? How do we assess their effectiveness? It can be difficult to internalise, to assess your own site signage. One method (highly recommended) is to ask someone who is NOT

[4] Of course, you must ensure that everybody wearing yellow coats have CRB check.

Table 11.1 Decision support analysis – communications

The Big Event	*Local Media*	*Regional, National Media*	*Websites*	*Social networks*	*Stewarding and security staff*	*Council Staff*	*Family and Friends*
Under 16s							
16–25 years							
26–35 years							
36–50 years							
50+ years							
Families with young children							

Key	Effective	Partially Effective	Not Effective

familiar with you site, who has never been there before, to walk round and comment. However, signage is only one method of communicating with a crowd, before, during and after an event. There are many dimensions, and a matrix approach lends itself to assessing the quality and effectiveness of your communications. Below is an example, where down the side, we have listed the different demographics, and across the top, the different methods of communication. We use colour (red/amber/green) to indicate effectiveness of that communication channel, with that demographic. This is design along the same schema as the DIM-ICE risk models.

You define this table as a group activity, where the focus is on the content, the discussion around the effectiveness. You learn from the process of completing the matrix (Table 11.1).

We used this approach in Saudi Arabia to evaluate the different methods of communicating with different nationalities.

1. Heads of Groups (Sheiks)
2. Service Groups
3. Mushideen
4. Screens
5. Internet
6. Travel/Local Companies
7. International Hajj Missions
8. Saudi Embassies in Origin Country

Not Effective	
Slightly Effective	
Effective	
Very Effective	

Table 11.2 Saudi Arabia – nationalities and effectiveness of communications

	1	*2*	*3*	*4*	*5*	*6*	*7*	*8*
African								
Turkish and European								
Southeast Asian								
South Asian								
Arabian								
Iranian								
Internal Hajjis								
Unregistered Hajjis								

Table 11.3 Risk profile by location and time

Australian Event	*11*	*12*	*1*	*2*	*3*	*4*	*5*	*6*	*7*	*8*	*9*	*10*	*11*
Pathways													
Bars													
Catering													
VIP													
Taxi													
P'd'ton Gates													
W'hra Gates													
Stage													
Toilets													

We developed a model (Table 11.2) during a course run in Saudi Arabia in 2004, planning for the Jamarat Bridge[5] project. The model (Table 11.3) comes from a course in Manley, NSW, Australia. Table 11.3 was used to assess security and police deployment based on experience of area of elevated risk/congestion.

CROWD BEHAVIOUR AND THE DIM-ICE MODEL

We can also apply a demographic classification to the DIM-ICE models. There are many different descriptions and models of crowd behaviour, for example, you could define a crowd mood/behaviour as casual, cohesive,

[5] https://www.gkstill.com/CV/Projects/Jamarat.html

Table 11.4 DIM-ICE model for different crowd types

Casual	*Ingress*	*Circulation*	*Egress*
Design			
Information			
Management			

Cohesive	Ingress	Circulation	Egress
Design			
Information			
Management			

Anti-social	Ingress	Circulation	Egress
Design			
Information			
Management			

expressive, anti-social, aggressive, incident, etc. Each different mood/behaviour may require different strategy or tactics for communication, and for management. You can model different crowd types in the DIM-ICE matrix. For example, an anti-social crowd may require a different crowd management strategy to a casual or cohesive crowd (Table 11.4).

This approach can help you identify are the crowd policing/stewarding tactics. You may require different strategies and tactics to manage each of these diverse types of crowd behaviour? By creating your own classification system, you will develop a vocabulary and identity for those strategies. Sometimes, the solution to such undefined characteristics as crowd mood may be assessed and understood through this kind of modelling. Using the matrix approach can assist in defining your strategies for your crowd management, for distinct types of crowds and how/when to change strategy if, or when, the crowd behaviour changes.

INTERPROFESSIONAL COMMUNICATIONS

Assessing how to communicate with the crowd is one aspect of our crowd safety and risk analysis toolkit. Interprofessional communications are equally important, how various agencies at an event, communicate with each other, specifically during emergency situations, also lends itself to this kind of matrix analysis. Table 11.5 is an example from one of our workshops, on how the various groups at a major event site, communicate. In words, the method of communication, and the colour adds the dimension

Table 11.5 Inter-professional communications

Comms Analysis	*EOC*	*Police*	*First Aid*	*Security*	*Ushers*	*City Health*	*Crowd*
EOC	Direct Contact	Comms officer in EOC	Comms officer in EOC	Ushers to go to EOC Channel	Via EOC Channel	Direct contact, in proximity	Via Staff and security, loud hailers, social media SMS
Police	Command post in EOC	Own System, established structure and protocols	Through EOC, direct contact with comms	Direct contact at local area, through EOC	Direct contact at local area, through EOC	Through EOC	Loud hailers, external, direct contact
First Aid	Comms officer in EOC	Direct contact in EOC with comms	Own System, established structure and protocols	Direct contact at local area, through EOC	Direct contact at local area, through EOC	Through EOC	Direct Contact
Security	Comms officer in EOC	Through EOC, direct contact locally	Dedicated channels, recorded through EOC	Dedicated channels, recorded through EOC	Direct contact at local area, through EOC	Through EOC	Direct contact, loud hailers
Ushers	Ushers go to EOC Channel	Direct contact at local area through EOC	Through EOC, direct contact at treatment rooms	Direct contact at local area, through EOC	Dedicated Channel, runners used	Through EOC	Direct contact, loud hailers
City Health	Direct Contact	Through EOC	Through EOC	Through EOC	Through EOC	Own system, direct contact	Through EOC
Crowd	Through staff and security, social media, SMS	Direct Contact	Through Staff, and security, direct contact at rooms	Direct contact. Social media, SMS, through EOC	Direct contact	Post event, email and phone	Social media, directing contact, limited to area, line of sight

Highly effective	Effective	Partially Effective	Needs improving

of effectiveness. This can highlight potential problems, during desktop analysis, issues that you may not discover until your commination problem occurs during an incident.

One example of interprofessional communication problems was the failure of radio channels during the response[6] to the Sept 11th bombing of the Twin Towers,

> Systems were variously destroyed or overwhelmed by loads greater than they were designed to carry or failed to operate as intended or desired.

EMERGENCY PLANNING

We have run courses and workshops on various client site around the world. During one course, the client presented us with their contingency planning documents. It was a massive document, well over two hundred scenarios in their planning. His problem was two-fold, they cannot memorise all the action/reaction contingency plans, and they might not have the time to look up a scenario *during an incident*. Clearly not a suitable course of action. Secondly, what if we have an incident that is not in the book?

We presented the matrix approach as a solution and expanded his options for contingency planning. We can, using a matrix, define locations down the side, and threat along the top. Again, description in words and the box in colour[7] (Red/Amber/Green) to define severity. In one template (see Table 11.6), we covered fifty-six different scenarios, as you can see there is scope for more columns, further scenarios. It highlighted the threats have greater impact, depending on their location on the site.

MANY APPLICATIONS

You can model four dimensions in this matrix approach. For example, location down the side, time along the top, colour to represent risk and an icon[8] to show the crowd dynamics. Over the years, we have seen different uses for the matrix approach, to support critical decisions, strategic and tactical, and to provide control room options. Again, the process of completing the matrix provides insight to the site issues. If you cannot record information into any of the boxes that tells you there is a

[6] https://en.wikipedia.org/wiki/Communication_during_the_September_11_attacks

[7] Although the tables in this book are in greyscale, the DIM-ICE model should be in colour. We did choose colours and their greyscale, to cater for colour vision deficiency in the population. You should check your users, to make sure the colour scales are interpreted as intended.

[8] https://thenounproject.com

Table 11.6 Threat, location analysis

	Cold weather/rain	*Hot weather*	*Crowd crush*	*Flooding*	*Bomb threat*	*Active shooter*
Green Gate	Negligible	Crowd Frustration, Overheating	Negligible	Possible	Possible Target	Likely Target
White Gate	Negligible	Crowd Frustration, Overheating	Negligible	Possible	Possible Target	Likely Target
Red Gate	Negligible	Crowd Frustration, Overheating	Negligible	Minor Possibility	Possible Target	Likely Target
Main Stage	Negligible, Reduced risk	Crowd Frustration, Overheating, Dehydration	Highly possible	Negligible	Highly possible, High Density, Crowd Crushing	Highly possible
Tent Stage	Higher Densities	Crowd Frustration, Overheating, Dehydration, passing out	Highly possible	Negligible	Highly possible, High Density, Crowd Crushing	Highly possible
Concessions	Negligible	Crowd Frustration, Overheating	Negligible	Negligible	Negligible	Negligible
Arena Entrance	Negligible	Crowd Frustration, Overheating	Highly possible	Negligible	Crowd Crushing, Hugh Density	Likely Target
Towpath	Negligible, River Level Monitoring	Crowd Frustration, Overheating	Highly possible	Possible	Crowd Crushing	Likely Target

gap in your knowledge. Colour indicating severity, effectiveness, words describing reactions or risks, two variables, this is a very flexible process of capturing a lot of information on a single page.

STRATEGY AND TACTICS

The technique lends itself to many different analysis techniques, such as police strategies and tactics, for various locations, and different crowd behaviours. Like any modelling tool, the more you practice with the matrix approach, the more you can get out of its application for event planning and control room information.

Chapter 12

Emergency situations

> If you fail to plan, you are planning to fail!
>
> Benjamin Franklin

The objective of using the crowd risk analysis tools is to *reduce* the potential for a crowd related incident. However, as Mark Breen outlined in his LaD paper (in the DIM-ICE chapter) there are limitations to preventing *all* threats to the crowd. In the UK, and USA, there are legal requirements to produce, and share, your emergency plan with the relevant local authorities. From the UK Government website[1]

> You must have plans in place to respond effectively to health and safety incidents and other emergencies that might occur at an event.
> This emergency plan should be in proportion to the level of risk presented by event activities and the potential extent and severity of the incident.

The outline on the UK Government website details the following steps:

- Consider the key risks
- Sharing your plans
- Develop an emergency plan
- Emergency procedures
- Have clear emergency roles and responsibilities
- Evacuation
- Show stop
- After the incident
- Testing and validation

We outlined the principles and applications of the risk assessment in a previous chapter. These requirements are similar around the world.

1 https://www.hse.gov.uk/event-safety/incidents-and-emergencies.htm

DOI: 10.1201/9781351053068-12

From the USA Ready.Gov[2] website,

> ...The first step when developing an emergency response plan is to conduct a risk assessment to identify potential emergency scenarios. An understanding of what can happen will enable you to determine resource requirements and to develop plans and procedures to prepare your business. The emergency plan should be consistent with your performance objectives.

As we have stressed, throughout this book and on our courses, you need to understand the principles and applications of crowd safety and crowd risk analysis if you are planning, or managing, places of public assembly. Part of that is the threat assessment and emergency plan. This is usually one of the key prosecutions questions we see in litigation. What was your plan? What was your "show stop" procedure? Did you implement your plan? From experience, there is an assumption that responsibility for crowd safety rests with the security/stewarding company. However, the event organiser will still bear the ultimate responsibility for the implementation of, or failure to implement, the plan. It is not a defence to state "we hired people to do that for us".

EMERGENCY PLANNING

We focus our work on planning, on the preparation and risk analysis phase of an event. Typically, to aid the desktop exercises conducted long before the event starts and before crowds arrive on site. Crowd risk assessment and crowd safety planning, when there is time to think, to evaluate options and consequences is invaluable with the right tools. The risk and safety insights, gained by using those tools, can then be utilised during the event.

We use the control room graphics to assist with incident detection and as an audit trail for the crowd ingress and egress. As such, we do not cover the topics related to emergency management. There are experts in this domain, for example, via the Institute of Civil Protection and Emergency Management[3] (Table 12.1).

Crowd science sits in the planning column. We designed the crowd risk analysis toolkit to help you anticipate risks, to assist in the preparation of your crowd management plan and your risk assessment documentation. Use them when you have time to think, time to plan. The methods we advocate are all low cost, such as pen and paper exercises. In other words, analysing the site, using the crowd risk analysis models, tools and techniques does not require capital investment or sophisticated technology. As we have demonstrated over the last few decades, we can teach the entire suite, and their applications in a short course.

[2] https://www.ready.gov/business/implementation/emergency
[3] https://www.theicpem.net

Table 12.1 Planning – Incident – Emergency Management

Planning	*Incident*	*Emergency Management*
Anticipate, Prepare, Risk Analysis, Proactive	Detect, react, inform the crowd	Rapid response, training, infrastructure, resources
Time to think, Time to plan, Low cost.	Fast reactions, time is critical	Time is critical, High cost

The aim of the crowd safety and crowd risk analysis tools are to identify those problems lying in wait before they manifest into significant risks to life and limb. Also, by using visual aids, specifically in the control room (as outlined in previous chapters), they can help identify a problem before it may develop into an incident.

INCIDENT DETECTION

During an incident, detecting a problem, reacting and then potentially informing the crowd requires a communication strategy. If we consider the incident cycle (Figure 12.1), anything that can reduce the time to respond, may reduce the scale of the potential incident. The timeline needs to be a short as possible. The problem is that as time progresses the crowd congestion can build up towards, around or away from your incident. Area can become congested.

You might have a brief time to respond to a situation that looks like it will develop into a crowd crush. CCTV investment, control room training, and understanding the expected crowd dynamics site are all investing in your event safety.

The consequences of an incident can run into hundreds of thousands of pounds/dollars. The potential damage to your reputation and your brand can be overwhelming. You should consider you investment in your safety

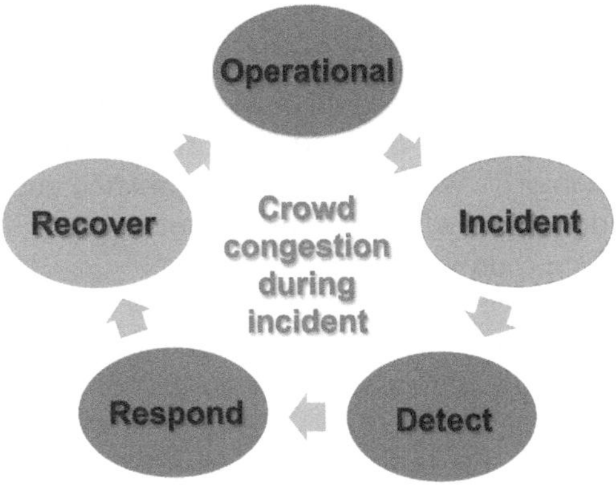

Figure 12.1 Incident cycle and crowd congestion.

systems like your event insurance. You hope you never need it, but when you do, demonstrating you have taken all due care and attention, can be the evidence you need to prove your case.

As we saw in the TED talk (Control Room Systems – Apollo Robbins[4] clip), it helps to place the information in front of you to ensure you are not referring to your memory of the previous, comparable situation. For example, you can use crowd density images to assist you in recording an audit trail, and as your early detection for potential overcrowding problems. This can be simple visuals for comparison checks or fill predictors. The purpose is to assist you in identifying situations before they become incidents, to reduce the time between incident and detection, and to act in a timely manner, to manage or mitigate overcrowding risks.

Where the cycle moves between detect and respond, that is the emergency response, element and, as we had stated above, not within the scope of the crowd sciences. However, it is essential that you consider emergency responses and have a solid contingency plan in place.

Consider the cost of training rapid response teams, reacting to the situation, infrastructure, roads, vehicles, these are all high-cost resources. We can see why investing in the planning stages may be the most cost-effective investment for an event organiser. To emphasise that point, in all of our expert witness cases, the failure of planning, the lack of an appropriate crowd management plan, the lack of appropriate risk assessment, the poor audit trails, the failure to understand the fundamentals of crowd safety have all proven to be the pivot points during litigation.

Being aware of crowd congestion, overcrowding during the incident, is an essential element for your crowd management plan. If anything causes the crowd to move away from any area, and migrate to another space, you need to consider how quickly congestion could build up. We have footage from events where a false alarm, such as an active shooter false alarm, caused by a sudden loud noise, or a fight breaking out in the middle of a crowd. There is a need to anticipate where the crowd may migrate, and if that migration will result in overcrowding.

Over the years, we have seen site issues resolved in desktop exercise, by the evaluation of "what if" the crowd moves to this point, area, location, with the inevitable question of "is there sufficient space for a safe mass movement to this area?". We have seen sites evaluated and then designed to avoid the potential of crowds migrating to blocked routes, dead ends and/or constraints along routes.

Understanding crowd risk analysis, and the activity cycle diagram of an incident, will help you develop a better understanding of both the crowd dynamics and the risk assessment process of incident management. Anticipating where the crowd congestion might build up is an important part of your emergency planning considerations.

4 https://www.ted.com/talks/apollo_robbins_the_art_of_misdirection?language=en

TORONTO – NATHAN SQUARE

At Nathan Square, Toronto, 2019[5], there was an incident where somebody within the crowd, discharged a firearm. The video footage shows the crowds migrating away from the site on one side of the event. On the other, officials were using the Public Address system and loudhailers, telling everyone to remain calm.

> The shooting happened as the Raptors sat onstage with Mayor John Tory, Premier Doug Ford and Prime Minister Justin Trudeau. The ceremony onstage continued despite the incident and people were being urged to remain in place. Officials at the rally urged the massive crowd to remain calm despite "an emergency situation nearby." When the rally eventually finished, the crowd dispersed normally. Speaking with reporters Monday evening, Saunders said there was a "tremendous amount of work" that went into keeping people safe at the festivities. "I can't describe the number of resources that you don't see that are involved in the type of work that is involved to get to where we are today," he said.

In the video footage of this incident, we can see two different crowd reactions. The different crowd behaviours were a result of types of communication systems used to inform and calm the crowd. One area had officials communicating, the other did not. When the crowd has little (or no) information it is like the domino effect, the crowd can react to itself, creating a wave of motion, rather than reacting to the incident. Which is why you need to consider your emergency systems. It may be vital that the crowd needs to know what is happening to not overreact and to respond appropriately.

You need to consider what information should be given, and if so, how much information should be given? How do you deliver it? How is it best expressed? And how do you monitor whether the crowds understand and act upon your information systems. You need to consider your event/site spaces; specifically, how the crowds may move into the various areas of your site. You should consider how various strategies for communication are implemented, received, understood, and acted upon? For example, is there conflicting information within your environment? Where and when might the crowd move?

SITE ANALYSIS

When asked to assess a site, we break large projects down in the following sections: Site Analysis – Scenario Planning – Implementation. First, we would evaluate the site maps and plans and outline the RAMP analysis and

[5] https://www.cp24.com/news/shooting-at-nathan-phillips-square-leaves-four-people-injured-sends-crowds-scrambling-1.4470268?cache=yes%3FclipId%3D373266

DIM-ICE risk matrix. Evaluate the occupancy, directions of ingress/egress and walk the site. In 2003, we modelled Canary Wharf, on request from the Metropolitan Police and Canary Wharf Management Ltd. The problems of mass evacuation from a multi-building site, following the 2001 Twin Towers incident, were prominent at every level.

Canary Wharf is a large multi-building complex surrounded by water. There are underground rail, over ground rail, busses, car parking and multiple bridges on and off the site. One of our projects was to assess the site, following the attack on the New York, Twin Towers (USA) and to analyse the site crowd evacuation strategies. This request came from the police, through the Canary Wharf Management group. The problem they encountered was dealing with the multiple false alarms and potential threat scenarios. At that time, the threat of aircraft attacks, sophisticated multiple vehicle attacks and the unknown were forcing the emergency services to radically rethink their tactics and strategies.

The emphasis in the guidance, the building codes[6], the civil contingency act and the local area regulations were all focussed on single building evacuations. However, a threat to multiple buildings led to the question – which group had the overarching responsibility to produce, and maintain, site wide plans? The complexity of a site, such as Canary Wharf, covered multiple local authorities. Transport for London (TfL), British Transport Police (BTP), the Metropolitan Police (the Met.), Canary Wharf Management Ltd (the Property Owners), each business, they all had an input.

The police also had a potential lability, in that to authorise a multiple building or site wide evacuation, unless there was an imminent threat to life, could result in significant financial losses to financial groups operating in Canary Wharf. The guidance[7] was reviewed and both the Police and the Canary Wharf Management realised a potential problem, in that they may be held liable for the financial losses incurred during an evacuation for a false alarm.

As with any large site, there are potential random threats in and around the area, from hoax calls to suspicious activities/vehicles. These threats could be located anywhere on the site, and to any scale. During previous evacuations, the police noted that the building occupants, when leaving the building, would call the police for further information. The circular nature of the influx of public requests for information, and resulting police response, is a distraction while dealing with a threat that could develop into a major incident.

We have collaborated with police forces around the world and collecting/ collating information from the public, specifically during the initial stages of an incident, can be source of confusion. For example, Berlin, Dec 2019, there were reports of an active shooter[8]. The public were calling the police

[6] https://en.wikipedia.org/wiki/Building_code

[7] https://www.gov.uk/government/publications/crowded-places-guidance/evacuation-invacuation-lockdown-protected-spaces

[8] https://www.express.co.uk/news/world/1222334/berlin-Starbucks-shooting-latest-Checkpoint-Charlie-Germany-Friedrichstrasse-Kochstrasse

with information about the location of the shooter, and from various locations. Trying to decipher and respond to all this information ties up police resources. Was it one shooter? Multiple shooters? With reported sightings at various locations it can stretch resources.

PROJECT ANALYSIS

These kinds of wide area projects followed a similar process. Firstly, the collection of the existing site evacuation plans. In this site, there are multiple businesses in single building; other buildings have a single organisation. When we compared plans, it was clear that the site required a multiple building evacuation strategy. The individual plans had created site conflicts. Building A's plan required staff to leave to the East, Building B's plan, which was on the East of Building A, required staff to leave to the West. Assembly points were within blast radius of a potential vehicle bomb. On the site, an underground mall and rail network connected buildings. Commuters would arrive via the underground and leave the same way. Never walking above ground. During a live evacuation test, it was apparent that many people had never walked off the site.

The second step is to analyse the geometry of the site, the network of routes, the egress route options the emergency vehicle approach routes and the crowd dynamics. We look at their existing planning measures and look at how they were implementing those and what we might need to do to change it for the future. So, during the initial site data capture, we would look at the sites strategic and tactical responses to a wide range of potential terrorist incident.

We use various site modelling tools on this scale of complexity. Typically, we would create a site route network, assessing the crow flow rates, along each route. Then site surveys[9], maps plans and diagrams are employed to evaluate the range of potential scenarios. What spaces can we use? How many routes do we have? What is the radius of various potential threats? How many buildings might be at threat, if a device were situated at point A, B or C?

Looking at their existing evacuation, planning and crowd management strategies and what scenarios might we need to deal with in future? On a large site we do not know where the threat might be, we do not know what the severity that the threat, small device or large device? This can be extremely complicated process to try and explain all the practical options in a document of a site emergency planning.

In a previous section of this book, we outlined using a decision support matrix. This allows the user to view multiple options for location, severity, responses and strategic responses. The matrix approach allows you to

[9] Where possible, always walk the site yourself.

document a wide range of options on a single table, but it lacks any geometry of a site, for referencing locations, which can be important.

In the second year, second term of the MSc programme, we set the students a task of producing a graphical version of the tools we developed for Canary Wharf. For that client, we solved the problems by using a grid-based network analysis (Figure 12.2), superseded in 2015, with a multi-location threat model.

The principle is that a coded message, the grid reference, and severity is transmitted, via intranet (internal network), mobile phone broadcast (via system called shop alert), via the building telephone network and walkie talkies. Everyone on the site, police, Canary Wharf Management, employees, fire Marshal's, would receive a weekly signal, they then click on the screen at that location, and then click on their building. We have pre-coded all the routing, capacity and severity radii into the programme, so that the screen displayed ONLY the safe, viable routes. Fire marshals were trained to lead the crowds away from the threat (all routes were calculated to provide crowds out and away AND first responders inbound) avoiding conflicts on the routes. The fire marshals adopt a pied piper approach. Everyone had a simple instruction – follow the crowd, keep going. When designing communication strategies, specifically for emergency situations, the information must be clear and concise. So, a training tool and a real-time tool in one.

The joy of teaching an MSc at this level is that you can demonstrate the principles, teaching the underlying mathematics of network analysis,

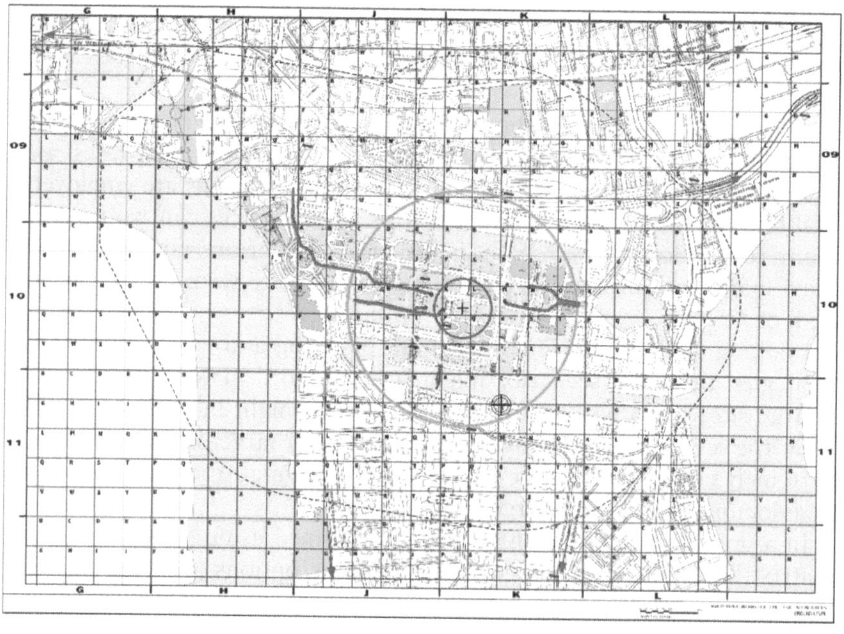

Figure 12.2 Canary Wharf Evacuation simulation/model.

crowd flow and the human factors, then ask the students to improve on those principles. They did. You can do all of this, without the mapping software, using an Excel spreadsheet.

USING EXCEL FOR REAL-TIME SITE EVACUATION MODELS

In simple steps, first import an image of your site into your spreadsheet. Make sure your site image has a scale. Then create a set of circles (to scale) for a 100, 200, 400 (metres or yards) as your exclusion zone. This might represent the blast radius of a small, medium and large explosive device. When you know the potential location of a threat, drag and drop the relevant exclusion zone over your map and your network. Move people away from the threat, along those lines. You will need to evaluate your communication strategy, and your red routes (for emergency situations) but the graphics of this approach make that problem much easier to evaluate during an incident.

In the image below, we show one of the MSc students projects, used for site planning, where (to scale) the circles, the map and the network are on a single worksheet. When report of a threat is received, the user drags and drops the relevant circle onto the map (top) and copy/pastes a copy to the network (bottom) (Figures 12.3 and 12.4).

This has proven to be remarkably simple to understand, create and use. In a control room, this gives you a quick graphical reference map of the site, the threat, its severity the routes in/out and, of course, is used by anyone receiving the grid reference and the severity, on the same excel spreadsheet.

We have seen versions of this principle, using a set of clear acetate circles, sticking to a site map, that you can slap on a printed map on a wall. It gives you a clear visual of which areas you need to clear. Trying documenting this, in a book of contingency plans, try describing this in words, exceedingly difficult, but in pictures, easy, effective and clear. As the saying goes, a picture is worth a thousand words.

ONLY FIVE STRATEGIC RESPONSES

In a previous chapter, we outlined the use of matrix analysis tools (based on Games Theory models) to assess strategies and tactics. When we need to plan for moving crowds away from areas of potential risk, using the routing models we need to consider other options. For example, an active shooter, it may be safer to keep people in their current location, to lock down. If we consider the wide range of strategy and tactics involved with threats, it seems like an unending range of potential threat, and hence potential responses. However, there are only five real options, five strategies for crowds (and individuals) during an emergency.

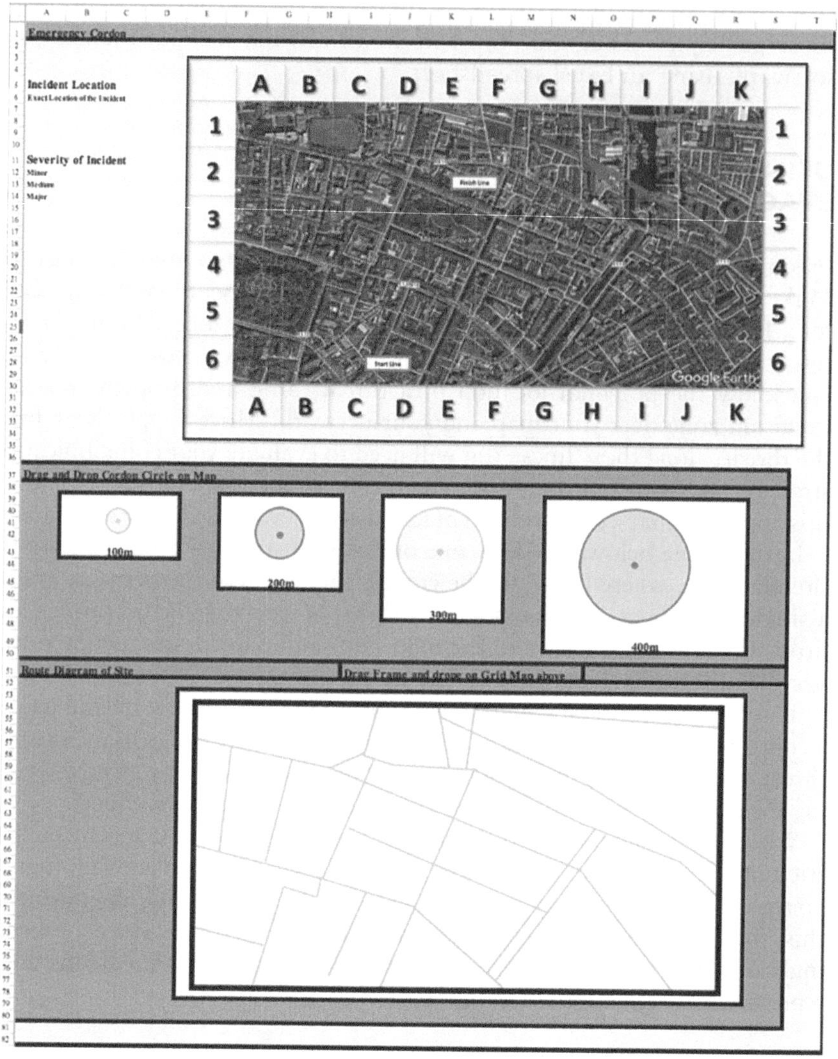

Figure 12.3 Town centre map in an excel spreadsheet.

1. Total evacuation

This is where we need to move people away from the threat. For example, a fire or suspect package. You can apply the GTFOT strategy (get the F*** out of there) where distance between the threat and the crowd is the important safety strategy. We have heard this called a FEAR response (F*** Everything And Run). If we define a total evacuation as clear the area. It does not matter which direction you go and just get away from the threat. Messaging is simple for this kind of strategic response.

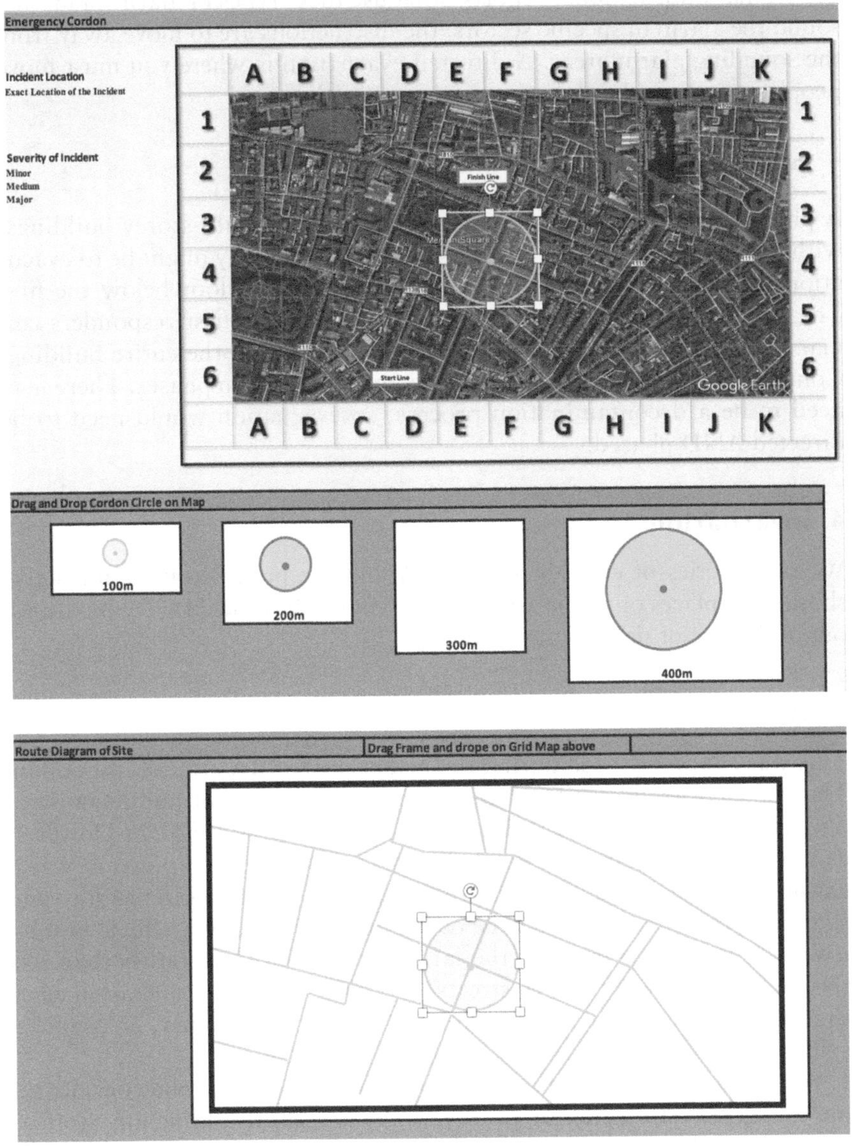

Figure 12.4 Using drag and drop to indicate severity on the map and network.

2. Directed evacuation

This is where we need to move the crowd in a specific direction, away from a threat. For example, we would not apply a Total Evacuation strategy if, say, there was a suspect package close to a main exit. This strategic response needs your messaging to be more complex. Consider a fire alarm, it has no directional information. Airports often use these types of alarms, but only sound the alarm in specific sectors, the instructions are to move away from the sounding alarm areas. A directed evacuation is where you must move people in a particular direction away from a threat.

3. Phased evacuation

A phased evacuation is the operating principle in multi-storey buildings, with appropriate fire suppression systems. The strategy might be to evacuation the floor of a fire, two floors above and one floor below the fire. The assumption is the fire suppression system, and first responders can manage the situation without the need to evacuation the entire building. Complex sites may plan to evacuate sector by sector, in phases. There may need to be a decontamination process, so evacuation would need to be directed AND phased.

4. Invacuation

We ran a series of courses in the USA and the police strategy for active shooters in places of public assembly is to move the crowd INTO buildings, out of the line of sight of the shooter.

5. Stay put

This was a strategy used during the height of the IRA attacks in London. The bombers would send coded messages to forewarn of imminent detonation. This gave the authorities time to evacuate a building. However, as most injuries occurred from fragments and glass, from pressure waves causing building damage along egress routes, it was often safer to stay-put, to use the building to protect the occupants. For example, vehicle bombs, it was often safer to move to the far side of the building rather than risk the crowd out and along open streets. Stay-put is also a consideration when the threat is outside, again this, might be an active shooter, or chemical spillage, or chemical attack.

You can create a matrix for these five strategic responses and consider the messaging and strategies for your site. The key to facilitating any strategy will depend on your communication/information systems. In the Canary Wharf project, we trained the Fire Marshals to be familiar with the various routes off the site. Hence the crowd instructions to “follow the marshals/crowd – keep going”. There is a small investment of time in planning for one of these five strategic responses but balanced against the risks of getting it wrong in an emergency, it is a small price to pay. In Table 12.2, we can insert the options for communication in the matrix, against the specific threat, and colour code the box for a level of severity.

Table 12.2 Communications matrix for threat against crowd strategy

Threat	*Total*	*Directed*	*Phased*	*Invacuation*	*Stay Put*
Fire					
Bomb threat					
Active Shooter					
Chemical					
Biological					
Radiological					
Other threats					

FURTHER RESEARCH

Soomaroo and Murray (2012)[10] studied 156 mass gatherings from 1971 to 2011 and looked at overcrowding, event egress and access points, fire safety measures, medical preparedness, emergency response as being the underlying causes for loss of life.

Although some of their assumptions and conclusions, are based on journalistic licence, and not major incident investigations[11], the principles of understanding causality, and hence defining both the risk assessment and the emergency response, training requirements, integrated crowd safety and emergency planning, are all sound conclusions. In our experience, it is the sites that do NOT approach planning for places of public assembly as an essential function, are the ones who simply do not consider that events might not always go to plan. That takes us back to Murphy's Law "Anything that can go wrong *will* go wrong".

When asked about the underlying causes of major accident and incidents, we would have to say that complacency is an underlying theme. Phrases such as

We've always done it this way …
We haven't had an incident yet …
We know what we are doing …
We've been in the industry for years …
We have lots of event experience …

These are typical comments I read in eyewitness accounts after an incident. Also, where organisers assume that "hiring a team" is the solution to all their site problems that the "team" has the responsibility for

[10] https://www.ncbi.nlm.nih.gov/pmc/articles/PMC3271949/

[11] For example, their assumptions of the cause of the crowd collapse at Ibrox (1971) is based on erroneous reports.

safety and then hire the cheapest bidder for the security contract or pay minimum wage. Another common issue we have seen is when organisers/promoters do not attend safety meetings, or simply do not understand the importance of crowd safety.

As we have already stated, you should treat your crowd management plan, your contingency plan and your safety plan in the same way as your event insurance.

Chapter 13

Summary

> A prudent man foresees the difficulties ahead and prepares for them; the simpleton goes blindly on and suffers the consequences.
>
> *Proverbs 22:3*

WHY DO WE NEED THESE COURSES?

At the start of our crowd science courses, we always ask how many of the delegates have, at the end of their event thought, "We were lucky that time". Usually most the delegates acknowledge with a sheepish "Yes".

We hope that one day the room will be silent, because successful crowd management is not about luck or chance. It is thorough planning, preparation, and risk management that you keep crowds safe in places of public assembly. Crowd safety management includes information, communication, and design. Crowd risk analysis includes understanding, mitigating, and managing crowd risks and understanding people.

Over 30 years ago, we noticed a problem in documenting crowd related information, and writing crowd risk assessments. The templates were all derived from static systems, such as stress loading on temporary demountable structures, where tolerances and breaking points could be tested to destruction. Were these templates appropriate for crowd dynamics?

We asked various authorities

> Why are you using this method for crowd risk assessment?

The typical response, a phrase that you should question, when you hear it was

> Because we've always done it this way.

Are you using the best method? Do you capture and communicate enough information for an appropriate crowd risk mitigation strategy? Is there a

DOI: 10.1201/9781351053068-13

better way to assess and document crowd risks? That is the fundamental motivation to develop a more comprehensive suite of crowd risk assessment tools, using visual techniques.

Our brains process visual content at an incredibly high speed. In fact, by one estimate, visuals communicate information 60,000 times faster than text[1]. However, this is a myth. Research by Matthew Dunn[2] would indicate that this is derived from the phrase, *a picture is worth a thousand words*, and that the mind can process an image 6× to 600× faster than the written word. As one popular meme states:

> Do not believe everything you read on the internet.
>
> Abraham Lincoln.

MORE EFFICIENT

If we consider that visual information is not language specific, that graphics do not require an ability to decode meaning from text, then images contain more information, understood by a wider audience, than the written word. Also, as the illustrations in this book demonstrate, even grey scale versions of the crowd risk analysis tools work.

We developed a visual set of tools for assessing risk, mostly through our international consulting projects, to help you understand the dynamics of crowd risks, and communicating those risks to an international audience.

USE THE RIGHT TOOL

There are five core risk models outlined in this book. As with any toolkit we do not need to use all the tools, all the time. Like any worker, we use the right tool for the right job. For example, we might use a DIM-ICE risk model to highlight risks that require future investment. DIM-ICE acts as a quality assurance marker, for tracking improvements, for communicating risks to a multiagency team. We might use a RAMP analysis to understand crowd movements approaching and departing the site/venue/event. You would use queueing models for ingress and egress to and from sites, the decision support analysis when you need to assess a multi-dimensional problem. Sometimes, we use all the tools, sometimes only partial application of one of the tools is appropriate.

---

[1] https://killervisualstrategies.com/blog/how-our-brains-are-hardwired-for-visua-content.html

[2] https://www.emailaudience.com/research-picture-worth-1000-words-marketing/

DIM-ICE

The DIM-ICE risk model is based on the research of past disasters and case studies. Designed to collect and collate information, then assembly it into a format (the matrix of boxes) that allows other to see the potential risks. As a communication tool, it serves to focus a multi-agency group on the issues around the influences on crowd behaviour (Design, Information and Management), the phases of crowd behaviour (Ingress, Circulation and Egress) and the modes of behaviour (normal and emergency).

This allows the user to outline the critical *elements* of a site/venue/event. We do not put an entire site in a DIM-ICE model (a mistake we often see when delegates use this for the first time). Do not write too many details in each box, negating its purpose. The DIM-ICE matrix is a shorthand version of the crowd management plan.

You should break your site into key segments and apply the model for sections of an event/site. It might be an entry point, an exit point, front of stage, junction or barrier system. For the DIM-ICE risk assessment you should ask what are the influences on the crowds moving in, moving around, and moving out of this space? What design information and management requirements would influence their behaviour?

CROWD DYNAMICS

Crowd dynamics relates to crowd density, the crowd flow, and of course, the spaces used by crowds; for example, static spaces, dynamic spaces and contingency spaces. We produce maps, plans and diagrams, for all three classifications of space. Where the crowds are static, or not moving, say spectating or viewing an event. Where the crowds moving and you may need to monitor or regulate crowd density, and look out for potential trip hazards Finally, what spaces might we need in the event of an emergency to move the crowd away from the threat, or hold the crowd until a threat is neutralised?

RAMP ANALYSIS

In the section on RAMP analysis, we looked at routes, areas, movement and people/profile of the crowd. For ***routes*** we create six diagrams, two routing diagrams for ingress (normal and emergency), for circulation (normal and emergency) and for egress (normal and emergency). For ***areas***, we outline the static, dynamic and contingency spaces. For ***movement***, we graph the arrival or departure profile over time (number of people up the side axis, and time along the bottom axis). For the people/profile, the profile of the crowd we would review similar sites. What do we know about the history or demographics of this crowd? What is our anticipated or expected behaviour of this crowd?

RISK AND CONGESTION MAPPING

Then we looked at risk and congestion mapping. This is where we map risk or congestion by location and time, showing location, duration, and severity of risk in a set of diagrams.

STRATEGY AND TACTICS

The decision support tools use another matrix approach to assessing an array of any one set of variables against another. It could be a risk against congestion, weather against communications, and crowd event type against congestion. There are multiple ways of using this approach. For example, determining the different responses and reactions for a threat at various locations around your site.

These tools are not simply academic research, they are practical tools, developed from real-world applications. They have had decades of use, not just by us, by over 10,000 learners that have passed through our courses (both face to face and online). The UK college of policing adopted these techniques for strategic and tactical decision-making.

Chapter 14

Crowd science essays

I cannot teach anybody anything, I can only make them think.

Socrates

In this chapter, we highlight delegates essay work from the short courses taught around the world.

When we teach, we challenge every student to assess the crowd, science models. If they agree, say so. If they can find fault, then provide an alternative, or augment the models accordingly, but support your option with evidence. Your opinion that "this should be added" is not science, explain why, demonstrate why your version works as an alternative, provide evidence that supports your model. When presented with a new model, such as Mark Breen's LaD classification, which enhances crowd sciences, we can adopt it. One example of this was the addition of "Invacuation" to the strategic alternatives for crowd evacuation alternatives.

We have seen other models, such as the USA "DIME-ICE", which fails to grasp that the elements in the Design, the Information and the Crowd Management *should* shape the crowds' expectations. Adding expectations as a column is counterproductive for two reasons. The first is that the model was based on the history of causality of accidents. It is used to view the specific area/section of the site through the lens of past disasters.

We do not see the evidence that "expectations" were a *cause* of an incident; instead of failing to manage, to shape the crowds' expectations was a distal cause. By adding expectations as a causal element, you shift the focus of causality towards the masses.

Expectations are within the context of the decision support models, where you outline your strategy and tactics for crowd management, where you assess how to shape the crowds' expectations. The second issue with adding expectations as a fourth column is that the model no longer fits into the catastrophe theory[1] model that we teach at Level 7 (Masters, MSc in Crowd Safety and Risk Analysis).

1 https://www.britannica.com/science/catastrophe-theory-mathematics

DOI: 10.1201/9781351053068-14

CATASTROPHE THEORY

Catastrophe theory is a branch of geometry where we depict the variables and resultant behaviours as curves or surfaces. The formal development of the theory is credited chiefly to the French topologist René Thom.

> A simple example of the behaviour studied by catastrophe theory is the change in shape of an arched bridge as the load on it is gradually increased. The bridge deforms in a relatively uniform manner until the load reaches a critical value, at which point the shape of the bridge changes suddenly—it collapses.

The catastrophe surface has three dimensions, and so does the DIM-ICE model. By adding columns – as the USA DIME-ICE, the model no longer fits in the catastrophe theory framework.

The same problem applies to a model called "DIM - ALICED – Design, Information, Management – Arrival, Last Mile, Ingress, Egress, Departure". This model may be useful for events where there is a last mile, but it combines the crowd dynamics and crowd risk in one matrix. These are different concepts and, as such, can only confuse the concepts of localised site risks and crowd dynamics across a wider area. The problem with these, and other derivations, is that they deviate from the primary objective of the DIM-ICE model, which was to collect, collate and communicate a risk assessment of a *specific section* of a site.

The aim of using different tools is to focus the user's attention that these are different perspectives on risk, and distinct types of analysis. Mix them all together and try and produce one model that does everything, and you are back to the same template currently in use.

We do not advocate trying to fit a too large an area, or a whole event into a DIM-ICE model. As we highlighted in a previous chapter of this book, the DIM-ICE application is multi-scalar, you apply DIM-ICE to sections of a site, then see the larger picture (see the Canada Day example in Chapter 2). Our aim was to produce a more versatile method of differentiating distinctive types of crowd risks for the built and complex environment, hence a toolkit for crowd risk analysis.

STUDENT PAPERS

As we mentioned in earlier chapters, the structure of this book follows the online VLE courses Level 3 – Fundamentals of Crowd Science course. We use the Introduction to Crowd Science course (Level 5) at the core theoretical element for the Minor at Breda University of Applied Science (BUAS). This started in 2019 and we have seen over 100 students pass through the course at BUAS. The first essay in this section is from the winner of the graduate year 2021 essay competition. We present four of the essays in this chapter, there are more in the online content and PDF version of the book.

Online content (https://www.routledge.com/Applied-Crowd-Science/Still/p/book/9781138626560)

MANON GIJBELS – EVENT SAFETY PLANNING

About Manon

> Manon Gijbels is a fourth-year International Tourism Management student at Breda University of applied science in The Netherlands. The tourism industry is fascinating to her, because of her own fondness for travel. During her last year of studies, she participated in a minor Crowd Safety in Hubs and Events because of an interest in understanding the causes of crowd incidents/accidents. Moreover, a willingness to understand the management that goes on behind the screens of an event. Once graduated, she is eager to start her career and excited to broaden her horizon. A future career goal being to work in the field of crowd safety and to contribute to a future in which "the safety of the people is the highest law".

An event safety plan should incorporate crowd science

I remember visiting Pinkpop Festival a week and a half after the Manchester Arena bombings in 2017 (Chan & Smith, 2017), which changed my outlook on visiting such a big event. My mother told us to be aware of anyone wearing a backpack and to look for emergency exits anywhere we were on the festival.

The subject Crowd Science and all its components like dynamics, behaviour and analysis are a fairly new and unknown topic to many. Most individuals visit a festival and are unacquainted with the "what ifs"' that roam the minds of crowd safety specialists. We live in a world where terrorist attacks and mass shootings have taken over the news, before COVID-19 happened at least, which scared people from visiting concerts, going on holiday or even to school (Cigna, 2016). Thankfully, there is an increasing number of specialists to help strive for a future in which "the safety of the people is the highest law" (Cicero, 45BC). Which is why my vision for the future of event safety is a customary event safety plan with strict guidelines that is based on the different risk levels, like in the UK (Health and Safety Executive, sd).

Crowd science can be defined as understanding the causality for crowd incidents/accidents. These incidents mostly transpire at concerts, sporting events and festivals, so places where a large crowd is present at a site. Most disasters occur because of a mismatch between the capacity of the area and the number of people trying to move into the space. These accidents are connected to either crowd-behaviour, dynamics or risk analysis (Still, 2019). Crowd simulations are also a tool used in crowd science. Most of which were reasonably foreseeable when they were still on the drawing table. Predictable is preventable.

Crowd simulation is the process of simulating the movement (or dynamics) of a large number of entities or characters (Thalmann, 2016). It allows for experimenting in a way that is not possible in real life, for obvious safety reasons. Moreover, it shows how risks develop into incidents and how these can turn into disasters (Vergunst, 2020).

Nevertheless, this computer-generated version is not the same as the real world. Simulations are very helpful and have added significant value to the world of crowd science, but they can be misleading. The person that is responsible for the simulations might make it look safe and feasible, but an expert need to be available to interpret the results. This is similar to the reviewing of event plans, if done by a person with little knowledge on the subject it can lead to dangerous situation (BBC, 2017).

The first element of crowd science is crowd behaviour which is defined as behaviour that is shown by individuals who gather in a crowd (Wijermans, 2011). It is difficult for experts to predict every factor that has an influence on crowd behaviour. Nevertheless, it is crucial that this behaviour is considered as one person's position, under certain key conditions, can have a substantial impact on the outcome of a situation. For example, a small fight that turns into a violent conflict could cause panic and insecurity among visitors.

Next, crowd dynamics can be described as the study of how and where crowds form, and move, above the critical density threshold of more than one person per square meter. The key theory of dynamics is the difference between low and high density. At low density (less than one person per square meter) people can move around freely and choose their routes and directions while reacting to their surroundings. High density (more than six/seven persons per square meter) could result in potential injuries and overcrowding. (Still, 2019) This is a crucial element in event planning as a significant number of incidents occur (Spence, 2020), because of the disparity between the area's surface and the amount of people trying to enter that space.

Following, a risk analysis is a process in which a risk is identified, and a value is assigned to that risk (Still, 2019), so recognizing dangerous situations when planning and assessing events. This crowd risk analysis is a concept that takes up a lot of time and effort as it must be done thoroughly to ensure a safe event.

Crowd risks in the event space are related to the static and dynamic risks. A static space is an area where the crowd is mainly stationary, and the opposite is a dynamic space where people are mostly moving (Still, 2019). This is important to know as starting point as it shapes the crowd's behaviour and how they will react to risks.

There are also three modelling techniques used for capturing and communicating event risk, such as a RAMP analysis, DIM-ICE meta-model and matrix analysis technique. The first mentioned instrument, a **R**oute, **A**rea, **M**ovement, **P**rofile analysis is a qualitative process that is performed while assessing and designing an event site both under normal and emergency circumstances. The next tool, a DIM-ICE meta-model (Design, Information, Management – Ingress, Circulation Egress) is used to collect information focused on crowd risks. The template allows the user to assess the behavioural influences of a crowd with an emphasis on the analysis during the different phases. Another technique to evaluate the effectiveness of a strategy is a decision-support analysis which allows the user to map one set of variables (weather) against another (bars) and colour the table to indicate the potential congestions. This technique helps identify how, for example, a change in the weather impacts the crowd.

Last but not least, a risk map can be created to highlight where congestions or queue forming might occur during a particular phase or at a certain area of the event (Still, 2019).

All these previously mentioned tools also use colour coding. The colour ***red*** is the high-risk indicator, ***orange*** portrays medium risks for the crowd and logically ***green*** means low risk. Colours are there to highlight potential risks and moments when extra attention is required (Still, 2019).

These aforementioned definitions all are quite fixed and generalised. Yet, the situations in which, for example, crowd dynamics are analysed are each influenced by external events such as weather changes, design of buildings and management styles. Moreover, these colours seem a bit vague and can be interpreted by the analyser in his/her own way. My opinion on a risk will differ from somebody that knows nothing or a specialist on crowd safety. This could lead to situations in which a risk is interpreted wrongly resulting in fatal incidents.

Also, as many organisers are not familiar and lack knowledge in the field of crowd safety, how can they be asked to draw up an event safety plan? Letting a person without sufficient understanding create a RAMP – analysis or DIM-ICE model would be seen as something to get done, or try to please the municipality, instead of a tool to ensure safety at their event.

There are sadly enough examples in which crowd injuries could have been prevented. One of these is the monster truck that crashed into the crowd during an "AutoMotorSportief" event which took place on the 28th of September 2014, in Haaksbergen, The Netherlands. A so-called monster truck, a self-built stunt truck of 1,500 hp with very large wheels, was supposed to drive over parked cars, at high speed, and turn back around but due to a fault in the relay the rear wheels did not steer, and it drove into the public. Three people died and almost 30 were injured in the incident. (RTL Nieuws, 2014) Safety was not the first and main priority here which resulted in terrible, preventable sufferings.

Event planning is something that takes up a lot of time and significant amount of money. Around the world, event organisers are planning more and more events while doing their best to minimise expenditure and maximise profit. Most organisers are familiar with the different contracts, building codes and marketing. Moreover, they know how many emergency exits are needed and how wide it needs to be but assessing the site for its unique elements and how these could mitigate risks is where organiser lacks in knowledge. Many of the safety event plans presented are simply a copy of last year's event because "why change if it worked out fine last time?" (Still, 2019).

Event safety plans are a requirement for licenced UK events (Health and Safety Executive) but is different per region in the Netherlands (Ferrari, 2019). The permit holder (usually the organisation of the event) is responsible for the order and safety of the participants and visitors on the event site. As any event, how small it may be, can involve risks. The event permit states which measures an organiser must take to guarantee public order, health and (fire) safety during the event and how incidents must be dealt with (Vrie, 2020). Also, a good safety plan can help when something should go wrong despite all precautions. The plan clearly describes what was done to prevent these risks and shows that the organiser has taken all precautions to prevent it.

Yet, the costs and time that need to be put into creating an event safety plan are an argument from the point of view of an event organiser. These businesses survive on profit and need to maintain cost controls to do so; safety does not add to the profits at all. Safety managers have different interests than event organisers do maximise safety versus maximise investment. The safety budget is always under pressure as it needs to compete with the goal of maximum returns. For example, simulations are a great way to get an understanding of crowd risks and enable one to experiment with behavioural assumptions. Sadly, the costs of a crowd simulation system are quite high and do not always fit in the safety budget (Still, 2019). Nevertheless, the consequences, for both visitors and the organisation, of a possible incident outweigh the costs of a safety plan by far.

It baffles me that these event safety plans are not a requirement in the Netherlands. As mentioned before, there have been enough incidents already and when are we going to learn? The arrival of national Handreiking Evenementenveiligheid (HEV) strengthened the work process at municipalities in 2012. They created a document that contains a description of the organisation and process surrounding events. It serves as a guideline for municipalities and safety regions. Regrettably, the HEV is not the national standard for work processes resulting in most municipalities carrying out these procedures their own way. The differences between cities, regarding work processes, are bigger than they should be (Schaap, GEMEENTEN MOETEN ÉÉN KADER GEBRUIKEN VOOR VEILIGE EVENEMENTEN, 2015).

To get back to Haaksbergen, the permit for the event was provided on a "routine" basis as it had been hosted seven times before, so it was handled as a normal regular application by officials. These officials did not seek advice from the police or fire brigade, as it is not common at such an event, according to the municipality (NOS, 2014). A risk assessment would have shown where dangers could occur. A RAMP analysis would have highlighted that the Area where the event took place was too small for a monster truck to do its' stunt safely. Also, the crowd was standing too close to the stunt and did not have enough space to safely get away in case of danger. Moreover, international guidelines from the Monster Truck Racing Association, an American and European organisation, state that the public is not allowed to be standing where the vehicle is going (Monster Truck Racing Association, 2014).

A thorough risk assessment would have shown the possibility of the event resulting in an incident was "reasonably foreseeable". The organiser, AutoMotorSportief, was negligent in taking the necessary precautions in drawing up a sufficient event safety plan. Moreover, failing to take the responsibility of guaranteeing a safe environment. The mayor (municipality) should have said no in the interests of public order, safety, public health and the environment. According to the committee, he made a wrong assessment of interests and did not act according to the rules. Furthermore, the requirements that were set for the event were inadequate and impossible to enforce (NOS, 2015).

Event safety plans are checked and approved by the organiser, host city and emergency services. The assessment of these plans is dependent on the subjective opinion of a police officer, fire marshal and municipality employee (Schaap, GEMEENTEN MOETEN ÉÉN KADER GEBRUIKEN VOOR VEILIGE EVENEMENTEN, 2015). These are qualified people who work these positions and are hopefully specialists in the work field.

The common basis is important in these assessments, if these people would all have the same framework and guidelines then the assessment would become less person-dependent.

My opinion is that the document provided by the HEV should be compulsory for every event organiser to provide to the host city. It would be a set document, no differences per region, at events. This would require organisers to analyse the whole event site and to zoom in on spots where possible dangerous situation could occur.

In the case of Haaksbergen, a permit was provided based on an application that did not even mention the word "monster truck" while this was part of the event all along (RTL Nieuws, 2015). A change in event should have, logically, resulted in a change in the event plan. But even if it was mentioned in the application, knowledge and expertise are necessary to recognise and manage special risks, such as with stunts. The licensing authority must be able to recognise these risks and propose appropriate measures. In many (smaller) municipalities, like Haaksbergen, this knowledge is not always available and there is a lack of a network to exchange this necessary knowledge and experience (Onderzoeksraad voor Veiligheid, 2015).

The proximate cause in Haaksbergen was that the relay of the rear wheels did not steer. There were also multiple distal causes like the crowd being too close to the stunt and the event site was too small for a huge monster truck to safely do the trick. Moreover, it was a design-related incident as the site proved inappropriate for its intended use. Long discussions can be held on whose fault it was, but the fact is that the event, in its current form, should have never taken place in such a small area (RTV Oost, 2014).

Crowd science and crowd safety can be better incorporated into event planning, licensing/permitting/approval and event delivery by having a customary event safety plan with strict guidelines. All event organisers should be obliged to participate in a course or workshop like the ones given by Keith Still (2019). Once all those people are aware of how easy it can be for a disaster to occur and how equally simple the solutions can be, it could contribute to a future of which crowd disasters are not part of.

My vision for the future of event safety is a set document with strict guidelines that is based on the different risk levels, like in the UK. Every organiser that wants to host an event should be obliged to deliver a safety plan to the different stakeholders within the city, such as the fire brigade, police and municipality. The document should include a thorough risk assessment with a RAMP analysis, DIM-ICE model and congestion maps, etc. It forces the organiser to reflect upon their event and look at it from a different angle, so instead of only looking at it from more of a visitors' point-of-view also from a safety perspective. These assessments are executed before the event takes place so instead of trying to patch up mistakes or looking at close calls after, these wrongs are righted beforehand.

There is still a great deal to be learned about the new topic of crowd science, such as with unique challenges like a pandemic and a six-foot society. Thankfully, we are learning from past experience, in which a mandatory safety plan can help, while trying to prevent those incidents from happening again. As "those who cannot learn from history are doomed to repeat it" (Santayana, 1906). We are striving for a future in which safety is the most important value, at events but also in the public space (06.01.2021, Gorinchem).

REFERENCES

(n.d.). Retrieved from Health and Safety Executive: https://www.hse.gov.uk
(2019). In G. K Still, Introduction to Crowd Science (pp. 2-5). Boca Raton, FL: Taylor & Francis Group.
Ahoy Rotterdam. (n.d.). Veelgestelde vragen. Retrieved January 8, 2021, from Ahoy: https://www.ahoy.nl/informatie/veelgestelde-vragen
Association, M. T. (2014, January 1). Veiligheidseisen International Monster Truck Racing Association. Retrieved from Instituut Fysieke Veiligheid: https://www.ifv.nl/kennisplein/evenementenveiligheid/publicaties/veiligheidseisen-international-monster-truck-racing-association
Bardy, M. (2020, May 25). COVID-19: Space Requirements & Crowd Density. Retrieved from Siflux: https://www.siflux.com/wp-content/uploads/2020/05/Covid19_space_density.pdf
BBC. (2017, September 19). Computer simulation. Retrieved from BBC: https://www.bbc.co.uk/bitesize/guides/zvxp34j/revision/3
Centraal Bureau voor de Statistiek. (2012). Babyboomers. The Hague: Centraal Bureau voor de Statistiek.
Chan, S., & Smith, R. (2017, May 22). Ariana Grande Manchester Concert Ends in Explosion, Panic and Death. Retrieved from The New York Times: https://www.nytimes.com/2017/05/22/world/europe/ariana-grande-manchester-police.html
Chapter 15: Mode Split. (2014). Retrieved from https://www.icpsr.umich.edu/CrimeStat/files/CrimeStatChapter.15.pdf
Cicero, M. T. (45BC). De Legibus. Roman Republic.
Cigna. (2016, March). Coping with Fear of Violent Attacks and Terrorism. Retrieved from Cigna: https://www.cigna.com/individuals-families/health-wellness/coping-with-fear-of-terrorism
Computer simulation. (2017, September 19). Retrieved from BBC: https://www.bbc.co.uk/bitesize/guides/zvxp34j/revision/3
Coping with Fear of Violent Attacks and Terrorism. (2016, March). Retrieved from Cigna: https://www.cigna.com/individuals-families/health-wellness/coping-with-fear-of-terrorism
Dijk, R.-J. v. (2020, October 27). Mobility. Breda, The Netherlands.
Dordrecht Marketing & Partners. (2020). Stadsplattegrond Dordrecht. Retrieved from Centrum Dordrecht: https://centrumdordrecht.nl/plan-je-bezoek/stadsplattegrond-dordrecht/
Dordrecht Marketing. (2021). Weekmarkt Statenplein & Sarisgang. Retrieved from Centrum Dordrecht: https://centrumdordrecht.nl/locatie/weekmarkt-statenplein-sarisgang/
Drama bij evenement Haaksbergen. (2014, September 28). Retrieved from RTL Nieuws: https://www.rtlnieuws.nl/nederland/artikel/1624806/drama-bij-evenement-haaksbergen
Drama monstertruck Haaksbergen: vergunning gemeente onterecht. (2015, October 28). Retrieved from RTL Nieuws: https://www.rtlnieuws.nl/node/893851
Esselink, R., Lenderink, T., & Nijenhuis, H. (2018, May). De Marktbezoeker. Retrieved from Centrale Vereniging voor de Ambulante Handel: https://www.cvah.nl/de-marktbezoeker/

Ferrari, C. (2019, March 22). Vergunning aanvragen voor je evenement – een handig stappenplan. Retrieved from eventbrite: https://www.eventbrite.nl/blog/vergunning-aanvragen-evenement-ds00/

Forsyth, D. R. (2014). Group Dynamics. Belmont: Wadsworth Cengage Learning.

Gebrek aan alertheid bij vergunning monster truck in Haaksbergen. (n.d.). Retrieved from Onderzoeksraad voor veiligheid: https://www.onderzoeksraad.nl/nl/page/6913/gebrek-aan-alertheid-bij-vergunning-monster-truck-in-haaksbergen

Groeneweg, J. (2020, May 4). Dordtse weekmarkt grotendeels goed verlopen: 'Maar sommigen zagen het toch als een uitje'. Retrieved from AD: https://www.ad.nl/dordrecht/dordtse-weekmarkt-grotendeels-goed-verlopen-maar-sommigen-zagen-het-toch-als-een-uitje~a0fe9eac0/

Hart van Zuid. (2020). Compleet vernieuwde busterminal Zuidplein. Retrieved January 8, 2021, from https://www.ahoy.nl/informatie/veelgestelde-vragen

Lieve, M. d. (2018, June 4). Bijna vier jaar na monstertruckdrama doet chauffeur zijn verhaal: "Kon ik het maar terugdraaien". Retrieved from NH Nieuws: https://www.nhnieuws.nl/nieuws/225741/bijna-vier-jaar-na-monstertruckdrama-doet-chauffeur-zijn-verhaal-kon-ik-het-maar-terugdraaien

Monster Truck Racing Association. (2014, January 1). Veiligheidseisen International Monster Truck Racing Association. Retrieved from Instituut Fysieke Veiligheid: https://www.ifv.nl/kennisplein/evenementenveiligheid/publicaties/veiligheidseisen-international-monster-truck-racing-association

NOS. (2014, September 30). Vergunning Haaksbergen "routine". Retrieved from NOS: https://nos.nl/artikel/704609-vergunning-haaksbergen-routine.html

NOS. (2015, February 17). 'Vergunning monstertruck ten onrechte verleend'. Retrieved from NOS: https://nos.nl/artikel/2019891-vergunning-monstertruck-ten-onrechte-verleend.html

Onderzoekcentrum Drechtsteden. (2020). Bevolking - Dordrecht. Retrieved from Onderzoekcentrum Drechtsteden: https://ocd.buurtmonitor.nl/jive?cat_open_code=c1

Onderzoekcentrum Drechtsteden. (2020). Monitor Coronacrisis. Sliedrecht: Onderzoekcentrum Drechtsteden.

Onderzoeksraad voor Veiligheid. (2015, May 20). Gebrek aan alertheid bij vergunning monster truck in Haaksbergen. Retrieved from Onderzoeksraad voor Veiligheid: https://www.onderzoeksraad.nl/nl/page/6913/gebrek-aan-alertheid-bij-vergunning-monster-truck-in-haaksbergen

Oozoo. (2020). Wetenswaardigheden, cijfers en statistieken over Dordrecht. Retrieved from Oozoo: https://www.oozo.nl/cijfers/dordrecht

RET. (2019, February 22). Cijfers metro: Rotterdam Beurs is drukste metrostation van Nederland. Retrieved January 8, 2021, from Treinreiziger: https://www.treinreiziger.nl/cijfers-metro-rotterdam-beurs-is-drukste-metrostation-van-nederland/

RET. (2019). Jaarverslag 2019 RET. Retrieved January 8, 2021, from https://corporate.ret.nl/media/2899/jaarverslag-2019-ret.pdf

RTL Nieuws. (2014, September 28). Drama bij evenement Haaksbergen. Retrieved from RTL Nieuws: https://www.rtlnieuws.nl/nederland/artikel/1624806/drama-bij-evenement-haaksbergen

RTL Nieuws. (2015, October 28). Drama monstertruck Haaksbergen: vergunning gemeente onterecht. Retrieved from RTL Nieuws: https://www.rtlnieuws.nl/node/893851

RTV Oost. (2014, September 29). 'Verantwoordelijkheid voor ongeluk Haaksbergen ligt bij gemeente en organisatie" 1200 kbps geselecteerd als afspeelkwaliteit Veel hulpverleners in Haaksbergen (Foto: Erik Schaddelee) Veel hulpverleners in Haaksbergen (Foto: Erik Schaddelee) Syan S. Retrieved from RTV Oost: https://www.rtvoost.nl/nieuws/199444/Verantwoordelijkheid-voor-ongeluk-Haaksbergen-ligt-bij-gemeente-en-organisatie

Santayana, G. (1906). Reason in Common Sense.

Schaap, S. (2015, June 10). GEMEENTEN MOETEN ÉÉN KADER GEBRUIKEN VOOR VEILIGE EVENEMENTEN. Retrieved from Centrum voor criminaliteitspreventie en veiligheid: https://ccv-secondant.nl/platform/article/gemeenten-moeten-een-kader-gebruiken-voor-veilige-evenementen

Schaap, S. (2020, September 10). Crowd Safety and Risk Analysis. Breda, The Netherlands.

Spence, S. (2020, October 28). Right-wing extremism: The new wave of global terrorism. Retrieved from World Economic Forum: https://www.weforum.org/agenda/2020/10/right-wing-extremism-global-terrorism/

Statista. (2019, September). Number of festivals in the Netherlands from 2013 to 2018. Retrieved from Statista: https://www.statista.com/statistics/673435/number-of-festivals-in-the-netherlands/

Still, G. K. (2019). Introduction to Crowd Science. Boca Rotan, FL: Taylor & Francis Group.

Thalmann, D. (2016). Crowd Simulation. In N. Lee, Encyclopedia of Computer Graphics and Games (pp. 1–8). Springer International Publishing.

van Driel. (2020, April 29). Markten in Zuid-Holland Zuid weer toegestaan, maar wel onder extra toezicht. Retrieved from AD: https://www.ad.nl/dordrecht/markten-in-zuid-holland-zuid-weer-toegestaan-maar-wel-onder-extra-toezicht~aa6efc74/

van Driel. (2020, April 30). Nieuwe regels tijdens de Dordtse weekmarkt: 'Houd afstand en blijf niet langer dan nodig'. Retrieved from AD: https://www.ad.nl/dordrecht/nieuwe-regels-tijdens-de-dordtse-weekmarkt-houd-afstand-en-blijf-niet-langer-dan-nodig~a5b28df5/

van Driel, I. (2020, April 29). Markten in Zuid-Holland Zuid weer toegestaan, maar wel onder extra toezicht. Retrieved from AD: https://www.ad.nl/dordrecht/markten-in-zuid-holland-zuid-weer-toegestaan-maar-wel-onder-extra-toezicht~aa6efc74/

van Driel, I. (2020, April 29). Markten in Zuid-Holland Zuid weer toegestaan, maar wel onder extra toezicht. Retrieved from AD: https://www.ad.nl/dordrecht/markten-in-zuid-holland-zuid-weer-toegestaan-maar-wel-onder-extra-toezicht~aa6efc74/

van Driel, I. (2020, April 30). Nieuwe regels tijdens de Dordtse weekmarkt: 'Houd afstand en blijf niet langer dan nodig'. Retrieved from AD: https://www.ad.nl/dordrecht/nieuwe-regels-tijdens-de-dordtse-weekmarkt-houd-afstand-en-blijf-niet-langer-dan-nodig~a5b28df5/

van Driel, I. (2020, April 30). Nieuwe regels tijdens de Dordtse weekmarkt: 'Houd afstand en blijf niet langer dan nodig'. Retrieved from AD: https://www.ad.nl/dordrecht/nieuwe-regels-tijdens-de-dordtse-weekmarkt-houd-afstand-en-blijf-niet-langer-dan-nodig~a5b28df5/

Vergunning Haaksbergen "routine". (2014, September 30). Retrieved from NOS: https://nos.nl/artikel/704609-vergunning-haaksbergen-routine.html

'Vergunning monstertruck ten onrechte verleend'. (2015, February 17). Retrieved from NOS: https://nos.nl/artikel/2019891-vergunning-monstertruck-ten-onrechte-verleend.html

Vergunst, N. (2020, June 2). Crowd simulation software visualises layout for social distancing. Retrieved from Utrecht University: https://www.uu.nl/en/news/crowd-simulation-software-visualises-layout-for-social-distancing

Voerknecht, H., & van Bokhorst, M. (2018, March). Benchmark actieve, gezonde, duurzame mobiliteit gemeenten. Retrieved from CE Delft: https://www.ce.nl/assets/upload/file/Rapporten/2018/4N74/CE_Delft_4N74_Benchmark_actieve_gezonde_duurzame_mobiliteit_gemeenten_DEF.pdf

Vrie, K. d. (2020, June 21). TIP! - Welke vergunningen heb ik nodig om een evenement te organiseren?. Retrieved from eventplanner: https://www.eventplanner.nl/nieuws/4569_tip-welke-vergunningen-heb-ik-nodig-om-een-evenement-te-organiseren.html

Westland, E. (2020, March 13). Druk op de markt na maatregelen corona: 'Iedereen wil wat extra's te eten in huis hebben' Met name eieren zijn vandaag niet aan te slepen op de markt in Kanaleneiland. Foto: RTV Utrecht / Evie WestlandMet name eieren zijn vandaag niet aan te slepen op de. Retrieved from RTV Utrecht: https://www.rtvutrecht.nl/nieuws/2027385/druk-op-de-markt-na-maatregelen-corona-iedereen-wil-wat-extras-te-eten-in-huis-hebben.html

Wijermans, F. E. (2011). Understanding crowd behaviour: simulating situated individual. Groningen: University of Groningen, SOM.

KRISTIN FINN – AN ANALYSIS OF CROWDS FOR EVENT SAFETY

About Kristin

Kristin Finn is a Lecturer in Event Management for the Department of Tourism and Hospitality at Plymouth University, and her main teaching roles are in crowd management, research and consultancy, and operational planning for events. Specifically, her research interests lie in matters relating to crowd behaviour, managing safety and the relationship between crowds and audience experience at events.

Prior to joining Higher Education, she worked in consultancy for almost ten years, most notably developing project evaluations, monitoring frameworks and organisational strategies for local government as well as visitor and audience behaviour studies for tourism, cultural and events organisations in the public and private sectors. This range of experience informs her teaching contribution and areas of professional interest.

Introduction

In 1995 Berlonghi observed that a crowd in a contemporary setting is a composite of multiple personalities. Thus, it can be argued that crowd identities must be studied in the plural sense (Borch, 2013), as heightened emotions can denote that one crowd can evoke a range of different behavioural patterns at different points during an event episode due to the unique characteristics of the individuals involved and the event-specific catalysts. This paper will explore the characteristics and catalysts commonly considered to have an impact on crowd behaviour in the context of events.

Knowledge of crowd typologies is important for the understanding of audience behaviour at events. Rutherford-Silvers (2008) identified five types of crowd (casual, cohesive, expressive, aggressive and explosive) ranging in the level of impact that they may elicit on an event and its audience. Similarly, Canetti (1973) identified six crowd types (invisible, bating, fleeing, prohibition, reversal and feast) and also used visual imagery to identify three main crowd symbols whereby he likened crowd movement to; the sea (dense and ever moving), fire (volatile with the ability to spread, destroy and behave expectedly) and the river (behavioural origins are taken more seriously than the goals). Building on this, Berlonghi (1995) identified a series of eleven crowd types. These ranged in their behavioural characteristics and the level of impact that they had on an audience. Core themes emerge across these typologies, which demonstrate a set of key behavioural profiles. Thus, it could be argued that crowd typologies must be understood by event managers to enable them to fully grasp the distinct profiles exemplified by common crowd types. It is also key for managers to understand how these profiles are influenced and the level of perceived risk that they denote in relation to the safety of those at an event in order to develop effective management strategies for such issues.

As already indicated, crowd behaviours can arguably be influenced by audience profiles and collective characteristics in a positive or negative way. It may also be said that threat to safety in crowded spaces can be impacted by a broad range of catalysts. For example, expressive or revelling crowds (Berlonghi, 1995) are often formed due to stimuli including event activities, performers' actions and the psychosocial factors of attendees. These can influence the audiences' experience in a typically positive way. The contributions of the experience economy and service design culminating in successful audience experiences is well documented in event management literature (Berridge, 2007; Hoffman et al., 2009; Mudie & Pirrie, 2006; Pine and Gilmore, 1999 & 2011; Rutherford-Silvers, 2008) and an expressive crowd is arguably an outcome of this approach. In such cases the event setting is designed to stimulate the senses and prompt specific behaviours with a desired effect creating feelings of positivity and revelry among the audience members.

It could also be argued that cohesive, spectator and participatory crowds, which are characteristics that straddle several of the common typologies, can also influence the audiences' experience in a positive manner.

Looking specifically at participatory crowds (Berlonghi, 1995), those engaged in running a marathon without incident are most likely to feel a sense of goal accomplishment and camaraderie resulting from taking part (Raj et al., 2009) and connecting them to the expressive crowd type (Rutherford-Silvers, 2008). That said, when security, social and man-made disaster factors (Berlonghi, 1995) combine in the form of terrorism threats at high profile events such as the Boston Marathon (BBC, 2012), the resulting outcome on the audience and the participatory crowd experience is undoubtedly extremely negative, which can influence a crowd to exhibit fearful behaviours. This links to the fleeing crowd type identified by (Canetti, 1973) presenting a high risk to safety.

There are a range of crowd types such as casual, thrill-seeker and political crowds (Berlonghi, 1995; Canetti, 1973; Rutherford-Silvers, 2008) that can produce positive behavioural outcomes but have the potential to evoke negative behaviour if managed inappropriately, triggered by either dissatisfaction or harm to the audience. Whilst demonstrator (political) crowds can produce positive shared experiences for the audience involved, catalysts such as police and security factors, and specifically, the use of excessive or unreasonable force can result in negative demonstrator behaviour, which can escalate to a more dangerous bating crowd (Canetti, 1973); this could be seen during the University of Birmingham Tuition Fee and Staff Wage Protest in January 2014, whereby a group of protestors who felt that they had been contained or "kettled" by police for four hours turned violent and were arrested on suspicion of assault and criminal damage (BBC Online, 2014).

Finally, the most negative and damaging crowd types to audience safety are deviant crowds which are potentially aggressive and hostile, rushing and looting, or violent and explosive, and also more cautious and fearful crowds which can generate crowd issues such as escaping and trampling and dense and suffocating incidents due to their need to avert risk and flee (Berlonghi, 1995; Canetti, 1973; Rutherford-Silvers, 2008). Whilst some crowd situations can be premeditatedly deviant such as the actions of football hooligans, those taking drugs and criminals engaged in opportunistic theft, with behavioural patterns influenced by factors including escapism, anonymity, tolerant attitudes and sometimes marketing (Brunt and Brophy, 2004), this is not always the case and some deviant crowd situations do not always start out in this manner. One example of this is the sensation seeking individual (Lepp and Gibson, 2007) characterised by motives of escapism in relieving frustrations and boredom, novelty, stimulation, disinhibition and boredom susceptibility (triggered by restlessness) which can become involved in negative and potentially deviant behaviours when this would arguably not have been their primary intention. Thus, those with a thrill-seeking profile may become deviant due to group identity and conformity (Postmes and Spears, 1998) or perhaps external catalysts within the event setting itself such as performer provocation or event activities and operational circumstances (Berlonghi, 1995) rather than other motivators.

It could be argued that there are links between the thrill seeking and deviant crowd types and Torgersen's "Impulsives" and "Hedonists" personality profiles (Hochwalder, 2009; Vollrath and Torgersen, 2002). Key common characteristics include that these individuals are often pleasure-oriented, undependable, attention-seeking, and in need of social confirmation, with little control over his or her emotional reactions, and appearing chaotic and changing. Equally, links can be made between the cautious crowd and Torgersen's "Insecure", "Brooder" and "Complicated" personality types (Hochwalder, 2009; Vollrath and Torgersen, 2002), where common shared traits include being self-conscious, dependent on others, overly sensitive to his or her own mental and physical experience, emotionally intense experiences with occasional outbursts, as well as being overly careful, orderly, shy, withdrawn and with a tendency to give up easily in certain situations. It would therefore seem that inherent personality characteristics can also play a part in influencing crowd behaviour.

What is frequently documented in the literature related to deviant and cautious crowd types is that the prevalence of these two very different, but equally negative categories often attract negative media attention for the event associated with them and can typically result in fear, harm or fatalities to the event crowd in attendance. One of the most documented outcomes of crowd-related incidents is the trample disaster, which can be caused by a range of catalysts. These include poor planning on ingress and egress. One example of this is the 2010 Love Parade stampede (BBC News Online, 2010), other examples may be caused by unexpected weather events such as the Indiana State Fair stage collapse (Tuohy & Ritchie, 2012). The aroused emotions amid the audience at the Pearl Jam and Roskilde Festivals (Shone and Parry, 2004) caused a similar type of behaviour and there are many and varied examples of this. Zen, Mao, Zhao and Yuan (2008) researched trample disasters by using the case study of an unrestricted, un-ticketed outdoor Chinese cultural celebration. In this study large crowds gathered on a bridge to watch the celebrations, this resulted in a crowd crush and a resultant bridge collapse. The study concluded that in most trampling incidents, asphyxiation is the main cause of death resulting from the crowd "leaning" force (when people lean forward or push on those in front) leading to suffocation rather than "trampling" as is often portrayed in the media. Trample disasters may occur due to a range of factors which include as venue failings (e.g., fires, capacity issues), equipment failings, weather events and security issues (Berlonghi, 1995), which in turn elicit panic in the crowd and may be exacerbated by alcohol or drugs (Tarlow, 2002). Crowd panic is linked to emotional arousal theory (Borch, 2013) and the resulting crush is the cause of many trample disaster fatalities.

Conversely, however, it could be argued that a strong social identity amongst those in a dense crowd, where similar situational norms and a shared bond with one another, make people more tolerant of a crowded situation. This owes its origins to those in the crowd perceiving this type of emotion to lower the risk to their personal safety and thus preventing them from exercising due caution. Often in such situations it is too late to act, and this outcome can be a contributing factor to the severity of the incident in highly dense and dangerous crowding incidents. This characteristic was noted in research by Alnabulsi and Drury (2014) around the effect of crowd density on safety at the Hajj, whereby though increasing levels of crowd density indeed reduced feelings of safety, this effect was moderated by the identification with the mass crowd and perceptions that others identified as Muslim.

As well as the more event-specific catalysts described above, links between crowd incidents and internal crowd warning signs (Tarlow, 2002) and also external destination-specific contributing factors (Abbott and Geddie, 2001) should be considered in the debate around crowd catalysts. According to Tarlow (2002) internally, the behaviour patterns of fan groups or followers can be considered alongside the prevalence of drugs and alcohol and a deviant, hedonistic party-fuelled atmosphere. Moreover externally, Abbot and Geddie (2001) suggest that event managers must consider the demographics of the host venue population, and the level of crime associated with the destination or venue in order to truly manage crowd incidents, although the two in many cases may not be related.

Thus, as Berlonghi (1995:242) stated, "disasters and many of the less serious problems associated with them will continue to occur if we prepare only for 'a large number of people gathered closely together'". Stott and Reicher (1998: 509) conducted research with police officers that showed that, despite a perception of crowd composition as heterogeneous, officers perceived crowd dynamics in a more classical light involving an anti-social minority seeking to exploit the mindlessness of ordinary people in the mass; police treating crowds in disorder as a homogeneous whole was argued to play an important role in escalating (if not initiating) collective conflict and was found to be a key component of social change in crowd contexts. Despite this call for concern almost twenty years ago, there is recent evidence to suggest that modern day policing strategies still prepare for undifferentiated methods and use of force when policing crowds and events (Hoggett and Stott, 2010; Stott et al., 2008). It is a holistic preparation delivery which is important in the context of crowds where both internal and external factors must be integrated to ensure that as many bases as possible are covered to mitigate against a disaster in a crowded space.

To conclude, whilst it has been established that crowds contain multiple personalities, the typologies and catalysts discussed in this paper are not only based on behaviour resulting from event-specific situations and scenarios but are also impacted by environmental ones and unique intrinsic behavioural patterns. These are borne out of a range of personal factors such as subculture and perceived norms. The challenges that exist for event managers relate to the manner in which crowd profile information is utilised in the management of risk and how this informs crowd management strategies.

REFERENCES

Abbott, J.L. & Geddie, M.W. (2001). Event And Venue Management: Minimising Liability Through Effective Crowd Management Techniques. *Event Management*. 6. 259–270.

Alnabulsi, H. & Drury, J, (2014). Social identification moderates the effect of crowd density on safety at the Hajj. PNAS. 111, (25). 9091–9096. [Online] Available at: www.pnas.org/cgi/doi/10.1073/pnas.1404953111 [Accessed 12.05.15].

BBC News Online (2010). Stampede at German Love Parade festival kills 19. [Online] Available at: http://www.bbc.co.uk/news/world-europe-10753399 [Accessed 10.02.14].

BBC News Online (2014). University of Birmingham protest: Thirteen arrested. BBC News Online: Birmingham and Black Country. 30th January 2014. [Online] Available at: http://www.bbc.co.uk/news/uk-england-birmingham-25955224 [Accessed 08.02.14].

Berlonghi, A.E. (1995). Understanding and planning for different spectator crowds. *Safety Science*. 18. 239–247.

Berridge,G.(2007).EventsDesignandExperience.Oxford:Butterworth-Heinemann.

Borch, C. (2013). Crowd theory and the management of crowds: A controversial relationship. *Current Sociology*. 0 (0). 1–18. [Online] Available at: http://csi.sgaepub.com/content/early/2013/07/10/0011392113486443[Accessed 18.07.13].

Brunt, P. & Brophy, K. (2004). English Seaside Resorts & the Deviant Tourist. *Acta Turistica*. 16, (1). 1–30.

Canetti, E. (1973). *Crowds and Power*. London: Penguin Books. Translated by Carol Stewart.

Hochwalder, J. (2009). Burnout among Torgersen's Eight Personality Types. *Social Behaviour and Personality: An International Journal*. [Online] Available at http://www.highbeam.com/doc/1G1-208133816.html [Accessed 02.09.13].

Hoffman, K.D., Bateson, J.E.G., Wood, E.H. & Kenyon, A.J. (2009). *Services Marketing: Concepts, Strategies and Cases*. United Kingdom, South Western Cengage Learning.

Hoggett, J. & Stott, C. (2010). Crowd psychology, public order police training and the policing of football crowds. *Policing: An International Journal of Police Strategies and Management*. 33 (2). 218–235.

Kemp, C., Hill, I., Upton, M. & Hamilton, M. (2007). *Case Studies in Crowd Management*. Cambridge: Entertainment Technology Press, Ltd.

Lepp, A. & Gibson, H. (2008). Sensation seeking and tourism: Tourist role, perception of risk and destination choice. *Tourism Management*. 29. 740–750.

Mudie, P. & Pirrie, A. (2006). *Services Marketing Management*. 3rd Edition. Oxford: Butterworth-Heinemann.

Pine, J. & Gilmore, J. (1999). *The Experience Economy: Work is Theatre and Every Business is a Stage*. USA: Harvard Business School Press.

Pine, J. & Gilmore, J. (2011). *The Experience Economy*. Updated Edition. USA: Harvard Business School Publishing.

Postmes, T. & Spears, R. (1998). Deindividuation and antinormative behaviour: A meta-analysis. *Psychological Bulletin*. 123 (3). 238–259.

Raj, R., Walters, P. & Rashid, T. (2009). *Events Management: An Integrated and Practical Approach*. London: SAGE Publications.

Rutherford-Silvers, J. (2008). Risk *Management for Meetings and Events*. Oxon: Routledge.

Stott, C., Livingstone, A. & Hoggett, J. (2008). Policing football crowds in England and Wales: a model of good practice? *Policing and Society*. 18 (3). 258–281.

Stott, C. & Reicher, S. (1998). Crowd action as intergroup process: introducing the police perspective. *European Journal of Social Psychology*. 28. 509–529.

Tarlow, P. (2002). Event Risk *Management and Safety*. New York: John Wiley.

Tuohy, J. & Ritchie, C. (2012). Indiana stage collapse caused by improper construction. The Indianapolis Star: USA Today.com. [Online] Available at: http://usatoday30.usatoday.com/news/nation/story/2012-04-12/indiana-state-fair-stage-collapse/54231792/1 [Accessed 10.10.13].

Vollrath, M. & Torgersen, S. (2002). Who takes health risks? A probe into eight personality types. *Personality and Individual Differences*. 32 (7). 1185–1197.

Wakefield, J (2013). Can Technology Help Avoid Stampedes? BBC News, Technology. [Online] Available at: http://www.bbc.co.uk/news/technology-24463736 [Accessed 24.10.13].

Zhen, W., Mao, L. & Yuan, Z. (2008). Analysis of trample disaster and a case study – Mihong bridge fatality in China in 2004. *Safety Science*. 46. 1255–1270.

ERIC KANT – INTRODUCTION TO CROWD SCIENCE

About Eric

Eric is the owner of Phase01[2] Crowd Management, a crowd safety consultancy in the Netherlands. Eric specialises in managing crowds in confined spaces and crowded places. He graduated from Manchester Metropolitan University with a master's degree in Crowd Safety and Risk Analysis (Crowd Science). Furthermore, Eric is a qualified agent-based modeller, certified by Lund's University. He has worked with Prof Still on many crowd counting projects, around the world.

Eric specialises in the various strategies that should be prepared for the emergency evacuation of large crowds and the associated effective crowd communication strategies. He is pragmatic and takes an evidence-based approach.

[2] https://phase01.nl

Eric has years of experience as an event safety specialist at the Dutch National Police where he provides expert advice to operational, tactical, and strategic commanders during the approval phase, preparation, and operational phase of major or high-risk public events, crises and mono/multidisciplinary operations. He was involved in hundreds of mass events: from national celebrations, concerts, and Christmas markets to Premiere league soccer fixtures and EDM festivals.

Introduction

In modern society crowds, or large concentrations of people, occur frequently and without serious problems. However, with the growing number of large – and increasingly larger events – the potential for incidents, accidents and mass fatalities are increasing. Crowd incidents, accidents and mass fatalities at large scale events or places of public assembly are continuing to take place. To reduce the incidence of such events, one has to understand the crowd and its behaviour. Berlonghi (1995) argued that it is a serious mistake to underestimate the crowd and that one must understand the different types of crowds that occur, "without this understanding, we are left with random attempts at crowd control and crowd management which may result in serious losses of life, health, property and money".

Crowd Science

Crowd Science can be described as the study of the underlying causes of crowd accidents and incidents (specifically mass fatalities in crowded spaces). The science is derived from the initial study on Crowd Dynamics (Still, 2000) which he describes "as the study of the how and where crowds form and move above the critical density of more than one person per square metre" (Still, 2000).

Key theories that underpin crowd science

Crowd dynamics: can be described "as the study of the how and where crowds form and move above the critical density of more than one person per square metre (Still, 2000)."

Crowd density: It can be defined as the number of people and/or objects in a given space (Drintewater and Gudjonsson, 1989). It is important to understand how and/or where crowds reach critical density and at what density there is a potential for a crowd crush. Fruin (1993) argues that a density of seven persons or more per square meter is dangerous: "When crowd density equals the plan area of the human body, individual control is lost, as one becomes an involuntary part of the mass. At occupancies of about seven persons per square meter the crowd becomes almost a fluid mass. Shock waves can be propagated through the mass sufficient to lift people off of their feet and propel them distances of 3 m (10 ft) or more".

Crowd behaviour: Fruin (1993) argues that crowd behaviour falls into two general categories, a flight response, or a craze. "Flight occurs where people experience either a real or perceived threat" and "a mass craze is a competitive rush to obtain some highly valued objective".

Flow rates: Still (2000) argues that "in confined spaces, crowd dynamics is a function of the local geometry, and this effect can dominate the crowd flow and behaviour especially during emergency egress".

Safety limits: Still (2000) states that "The effects of geometry (dead-ends, barriers, constriction in corridors and changes in corridor widths) create their own dynamic compressions in a moving crowd. We need to know the limits for safety and how close any particular environment is to those limits".

Risk assessment

If one wants to prevent crowd incidents or accidents from taking place, one must identify risks and assess these risks in an assessment. Risk assessments are mandatory in most European countries; it is unclear however if proper risk assessments are always conducted when they should be.

In this example, in the Netherlands, the "conventional risk assessment" is the most used method to assess risks.

Conventional risk assessment process

In a conventional risk assessment user assign a value to a defined risk and a value to severity. Users then multiply these values. The overall score indicates what action (a qualitative assessment) should be taken to control the risk.

The risk assessment example (Tables 14.1–14.3) shows the risk "Person receives a minor injury". The likelihood is 3 (possible) multiplied by severity 2 (minor)=Risk factor 6. According to the Risk Likelihood – Severity Matrix in Table 14.1 creating a risk with a "medium priority".

If a person gets injured in a "non-crowded" space on an event site, the severity can be "minor". However, the consequences of a person being injured in the middle of a narrow, crowded corridor during an emergency evacuation

Table 14.1 Conventional numerical risk assessment

Likelihood	*Severity*
1 = Rare 2 = Unlikely 3 = Possible 4 = Likely 5 = Almost certain	1 = Insignificant 2 = Minor 3 = Moderate 4 = Major 5 = Catastrophic

Table 14.2 Risk likelihood – severity matrix

Likelihood						
Severity		1	2	3	4	5
	1	1	2	3	4	5
	2	2	4	6	8	10
	3	3	6	9	12	15
	4	4	8	12	16	20
	5	5	10	15	20	25

Risk Likelihood – Severity Matrix as used at Dutch National Police, Regional Unit Rotterdam.

Table 14.3 Risk assessment

Hazard	*Persons at risk*	*Risk factor*	*Measures*	*Responsible*
Person receives a minor injury	All	3×2=6	Ensure there are adequate qualified first aiders on site during the event	First Aid Site-Supervisor

Risk assessment example taken from City of Unley (Australia) website.

are higher. Thus, the severity may change to "major" or "catastrophic" which would migrate the risk factor to one of "high priority" or "urgent action".

One can also discuss the values that are given to risks and severity. If we invite three participants to assign such values, we will invariably attain three different outcomes as such outcomes are based on personal preference and perception.

One can deduct from this approach that conventional risk assessment does not promote the delivery of descriptive values. Users are forced to describe the apparent risk in a short and concise value judgement which can easily be misconstrued by others.

Still (2013; 139–142) describes this process as highly subjective and flawed because of the condensed description. He identifies that important information about the risks might be lost when decoded or translated into a risk management plan.

Still further argues (Still, 2013: 63) that the conventional risk assessment process is "fraught with the potential for misinterpretation and misrepresentation for two reasons. The first is that crowd risks are dynamic in nature and one number[3] may not be sufficient to identify the risk and assign an appropriate outcome; the second is that in the risk assessment process, a list does not lend itself to description and hence the understanding of the dynamic nature of an event".

[3] A value that is assigned to a risk.

This paper argues, in a conventional risk assessment, that risks are recorded as a single line sentence which is not sufficient for a full and frank understanding of the identified risk itself. Important information about the risk may be lost in translation or transference owing to the condensed nature of the format. The person/persons that has the task of converting these risks into a crowd management or operational plan may miss important details about the identified risk which may lead to a flawed plan. The conventional risk assessment does not lend itself to the recording of the dynamic nature of risks and neither to the recording of detailed information. Owing to these flaws it can be determined that the conventional risk assessment is not an appropriate method for assessing risks.

Appropriate risk assessment methodology

Crowd risks are dynamic in nature and may change over time. Crowd risks are related to location, duration, and severity, meaning that the location of the risk can change over time from –for example – entrance gates to front of stage, that a risk can occur for just a short moment or for a long period and that severity can be different depending on many external factors or influences. Thus, a more appropriate methodology for determining risks as a fluid, dynamic assessment structure which changes dependent on those identifying, assessing and managing the risk at any given time.

Influences on crowd behaviour

During his Crowd Dynamics PhD research Still developed a database and began to categorise major crowd disasters. Still (2013: 78) "defined three fundamental elements that were common to major incidents:

1. Design related issues, where there appears to have been a miscalculation relating to crowd flow and the space capacity.
2. Information related issues, where there appears to have been an external influence on the crowd behaviour that induced an accident/incident.
3. Management related issues, where there appears to have been one or more management decisions that were responsible for the accident/incident".

Phases of crowd behaviour

We can break down an event in three primary phases:

1. Ingress, the crowd approaches the site or venue and passes the entry system.
2. Circulation, where the crowd moves around at the site or venue.
3. Egress, where the crowd leaves the site or venue.

Modes of crowd behaviour

Finally, there are primarily two types of behaviour, these focus on how crowds will react: during normal or emergency situations.

DIM-ICE matrix

Still (2013: 118) designed a template "that focusses the user's attention on the time-related element of crowd risks", "by structuring an event into times". In the template, we can outline the influences (design, information, management) (DIM) against the phases (ingress, circulation, egress) (ICE) under normal and emergency situations.

Still (2013: 120) argues that "the matrix approach not only focusses attention on key elements of time related risks" but that the model also "highlights what most crowd management plans fail to address: that incidents can occur during ingress, mid event and during egress".

Table 14.4 is an example DIM-ICE matrix. Colours can be applied to highlight the severity of the risk: Green=Low Risk, Amber=Moderate Risk, Red=High Risk. If we take a close look on the matrix, we see that there is an identified moderate risk in design during ingress: *Limited routes/ Congested spaces, no queuing design.* It is obvious that this risk might become a high risk during an emergency.

Table 14.4 DIM-ICE matrix example

Normal	*Ingress*	*Circulation*	*Egress*
Design	Limited routes, congested spaces, no queueing system	Limited routes, no queueing design	Limited routes, congested spaces, no queueing design
Information	Live screens, stewards	Routing towards areas and facilities, live screens	Live screens, stewards
Management	Progress monitoring procedures	Progress monitoring no protocol, chain of command	Progress monitoring no protocol, chain of command

Emergency	*Ingress*	*Circulation*	*Egress*
Design	Limited routes, congested spaces, no queueing system	Limited routes, no queueing design	Limited routes, congested spaces, no queueing design
Information	Megaphones, security, live screens, alarm, basic signage	No crowd congestions warning signs, measures	Megaphones, security, live screens, basic signage
Management	Progress monitoring procedures	Progress monitoring, chain of command	Overcrowding measuring is vague

Example DIM-ICE Matrix provided by Prof. GK Still at the short course Crowd Safety & Risk Analysis, November 2015, Utrecht - Holland.

As the DIM-ICE Meta model is designed to outline the three key influences (design, information, management) against the key phases (ingress, circulation, egress) at normal or emergency, one can argue that this matrix approach has more information, in a more accessible format, than a conventional risk assessment.

Legal, regulatory and risk management framework

In the Netherlands, licensing officers from the fire-department are responsible for the assessment of event emergency evacuation plans. For crowd flow these officers use the norms as defined in the Bouwbesluit 2012[4] article 2.108. Most important norm defined for evacuation is the capacity of an escape route or (emergency) exit, which is calculated at 90 persons per meter width per minute.

On April 1st, 2014, the Dutch foundation NEN[5] published the "NEN norm" NEN-EN 13200-7:2014: "Spectator facilities – Part 7: Entry and exit elements and routes" – this "specifies safety and design characteristics of entry and exit elements of passage that are used in spectator facilities, either singularly or in combination, to provide a route". According to NEN, the flow capacity is not more than 82 persons per meter per minute: "The maximum rate of passage in a level passageway shall be taken as not more than 82 persons/m/min" (NEN-EN 13200-7:2014: 18). Despite this new guidance, it is unclear if the licencing officers from all fire departments use it when assessing event evacuation plans. The problem with this guidance is that it has no legal basis for fire departments.

In the Netherlands, there are no official guidelines that define safe crowd densities. There are many documents available online (without any legal basis!) that nevertheless are used by several fire departments. These state that 2.5 person per square meter is the maximum density of a crowded space.

The Netherlands does not have the luxury of a "Green" or "Purple" guide, as is available in the United Kingdom. However, at the Dutch foundation "Stichting Evenementenhandboek"[6] is reproducing and translating the Purple Guide for use in the Netherlands under Dutch conditions.

In the Netherlands there are three types of events identified: "Category A": regular events; "Category B": events that need attention; "Category C": risky events. The distinction is made based on a "risk scan". This comprises an excel worksheet with dozens of variables and calculations showing A, B or C (as identified above) when filled in. It is mandatory to conduct a risk assessment in a category C "risk event" case. As mentioned before, it is unclear if appropriate risk assessments are always conducted when they should be.

4 'Building act 2012'.
5 NEN supports the standardization process in The Netherlands.
6 http://evenementenhandboek.nl

Crowd Science and its application within the events scenario

Currently there is a lack of appropriate knowledge in the field of crowd science. One tends to wait till something happens, learn from these mistakes, or not, as the case may be then change procedures, guidance, laws and plans. It is important that all parties involved in the safety, planning and licensing process understand the nature of crowd risks. If not properly understood we are condemned to experience problems that are, in fact predictable, and thus avoidable.

To reduce risks and to prevent crowd incidents and accidents from happening, it is importance to increase the awareness of this area and to invest in in the training and education of those responsible for the identification, assessment, analysis and management.

REFERENCES

Drintewater, J.A. & Gudjonsson. G.H. (1989). The Nature of Violence in Psychiatric Hospitals, in K. Howell S. Hollin C.R. (eds) *Clinical Approaches to Violence.* Chichester, Wiley.

Fruin, J.J. (1993). The causes and prevention of crowd disasters. Originally presented at the First International Conference on Engineering for Crowd Safety, London, England, March 1993. Revised exclusively for crowdsafe.com, January 2002.

Still, K. (2000). Crowd Dynamics. PhD Thesis presented to the University of Warwick, 2000.

Still, K. (2013). Introduction to Crowd Science. Boca Raton, FL: CRC Press.

Bouwbesluit[7] (2012): Artikel 2.108. Capaciteit van een vluchtroute[8]

http://wetten.overheid.nl/BWBR0030461/Hoofdstuk2/Afdeling212/2121/Artikel2108/geldigheidsdatum_02-12-2015

Wet Veiligheidsregio's (2010): Artikel 10: Het inventariseren van risico's van branden, rampen en crisis[9] http://wetten.overheid.nl/BWBR0027466/

NEN (2014): NEN-EN 13200-7:2014: Spectator facilities - Part 7: Entry and exit elements and routes https://www.nen.nl/NEN-Shop-2/Standard/NENEN-1320072014-en.htm

KEVIN KENNEDY – CROWD SCIENCE IN CANADA

About Kevin

Kevin Kennedy is a Staff Sergeant with the Ottawa Police Service (OPS) and has over 36 years of policing experience, including 15 years with the Ontario Provincial Police (OPP). While working with the

[7] 'Building act'.
[8] 'Capacity of an escape route'.
[9] "Inventory risks of fires, disasters and crisis".

OPP he was a member of the East Region Emergency Response team from 1993–1998. One notable assignment included serving as unit leader for the Public Order Unit, which deployed to crowd situations in large demonstration-type environments. With the OPS, Kevin has worked within the Event Planning Section since 2009 and was promoted to his current position as the section supervisor in 2015. He currently leads a team of event planners that prepare police operational and deployment plans for large and medium scale events in Canada's National Capital. These events not only include festivals and celebrations but also large-scale demonstrations. Kevin has been the lead planner for several of the largest events held in the city that include the National Canada Day Celebrations, the National Remembrance Day Ceremonies, Bluesfest and the National Capital Marathon. He was also the lead planner for OPS's 2017 Event Planning Team for large-scale city events celebrating Canada's 150 Anniversary. Many of these events attracted tens of thousands of people congregated in spaces not designed for that magnitude of spectators.

In advance of Canada's 150th, Kevin recognised that there was a significant need for better awareness and education surrounding crowd safety. In 2015 he began researching crowd safety practices and subsequently discovered the theories and educational materials developed by Dr. Keith Still. He initiated arrangements for Dr. Still to conduct a three-day short course in Ottawa in 2016 for police, emergency and other city service personnel. Following the short course, he submitted two written papers to Dr. Still that were graded "With Distinction". Only a few Canadians have completed this requirement and he was the first to achieve this goal and to merit the honour of distinction.

In 2017 Kevin was instrumental in making the arrangements for Dr. Still to return to Ottawa to present a second short course. Prior to that session Kevin was provided the Train the Trainer Course – once again, the first Canadian to achieve this accreditation. As a result, he delivered some of the course material during the 2017 session and is planning on engaging in more training sessions in the future.

In 2019 Kevin began his Master of Science in the field of Crowd Safety and Risk Analysis with Manchester Metropolitan University. He has recently completed his second year of study and looks forward to submitting his Dissertation in the spring of 2022.

Kevin continues to advocate for the consideration and implementation of enhanced crowd safety measures for all events. While most of the previous training in Ottawa has been provided to police and service delivery agencies, his focus for future courses will be geared towards event organisers, believing that crowd safety is the responsibility of all engaged in an event, and is not restricted to the emergency personnel or support services.

Introduction

Historically speaking, crowds have always been a part of human existence. Whatever the reason, people have been drawn together with the expectation that in numbers, their personal safety would not be at risk. In fact, it has even been reported that people with like-minded purpose feel safer as the group size increases (Alnubulsi & Drury, 2014). However, a feeling of wellbeing is in no way an assurance of overall safety. There are many other factors that prove to be hazardous including, but not limited to, the physical characteristics of the venue, the nature and history of the event and the organisers' ability to address planning considerations. Each of these factors can impact the density of the crowd.

It has been suggested that density affects the psychology of both the individual and the crowd itself. Applying this understanding during the planning stages of an event may prevent loss of life, injury or property damage resulting from unsystematic attempts at crowd management (Berlonghi, 1995).

Definition: What is Crowd Science?

Crowd Science moves beyond the simple definition of a crowd and studies the effects of the density and dynamics on the psychology of the individual and the group (Drury, 2015). Early studies focused on explanations for the transition from individualised thought to the "mob mentality". Crowd science moves a step further in studying how crowd density impacts crowd dynamics as a result of individual and crowd behaviour.

Key theories

The study of human behaviour in crowd situations is not simply an effort without purpose. "It's about understanding people's reactions and decisions when they're part of a group and learning to predict how they might move in any given situation" (Still, 2016).

Proxemics

In the 1950s and 1960, Edward T Hall, a cultural anthropologist, promoted his term Proxemics to define his "study of how humans being communicate though their use of space" (Hall, 1963). He identified four zones surrounding each person (public, social, personal and intimate). Hall suggested a person's behaviours are directly affected as a result of interspatial interactions and intrusions since we are all territorial by nature (Hall, 1989).

Crowd Density as it relates to Crowd Dynamics

Crowd Dynamics can be defined as the study of the how and where crowds form and move above the critical density of more than one person per square metre (Still, 2016). The dynamics will change as the density changes. Speed, flow rates and even sustainability are negatively affected by increased crowd density (Still, 2016).

With regards to legal issues (described later in this paper), it is important that one understand crowd dynamics as it pertains to the geography of a given site, in an attempt to reduce risks and promote overall safety.

Crowd behaviour

LeBon (1896) defined a crowd simply as a gathering of individual; whatever the chances that have brought them together. He suggested however that the psychology of a crowd can differ significantly from that of the individuals of which it is composed of. He stated that on some occasions the sentiments and views of the crowd can become the common mindset. LeBon proposed that the individual gives up his "conscious personality" adopting instead the group mentality. He identified it as an organised crowd, "a single being...subject to the law of the mentality unity of crowds" (Le Bon, 1896).

Social identity approach

The social identity approach recognises that each individual has his or her own personality, however it also identifies that the individual has multiple social identities that dictate the way that one interacts with others in group settings. In this regard, identifying with a group in a crowded space setting provides a sense of security that would not be present if one did not identify as part of the whole (us vs. them). This could, in essence, impact how a person behaves in terms of trusting the group dynamic. It is possible that this could lead to conditioned behaviour, disinhibition and/or the occlusion of personal standards to remain as part of the group (Reicher, n.d.).

Legal issues

The Criminal Code of Canada is the primary legislation that deals with negligence when failing to properly plan or consider the ramifications of overcrowding for events.

Section 220 refers to criminal negligence resulting in death while Section 221 deals with those resulting in bodily harm. Section 219 defines criminal negligence as

(a) doing anything, or
(b) omitting to do anything

which shows wanton or reckless disregard for the lives or safety of other persons.

Section 219 applies to "everyone" and includes organisers, planners and responders. The burden of proof falls to the Crown Attorney in showing that there was a marked departure from the standard of a reasonable person and in most cases, this is an overt act that is easily established. However, intention and deliberation are not solely necessary. In *R v. Sharp* (1984) indifference was found to be sufficient for establishing negligence. Therefore, it is incumbent on all planners and organisers to ensure that crowd safety is addressed and maintained in all stages.

Beyond the criminal aspect, civil courts, particularly in the United States, have heard a myriad of cases involving injuries that were sustained at crowded events. While there are no notable cases within Canada, the US cases focused on "duty of care". If Canada had a similar disaster, failure to address potential issues would leave organisers and corporations at risk of civil litigation.

How it can be incorporated into event planning

There are important elements that must be addressed to ensure the safety of a crowd. They include being aware of the social identity of the crowd (social rules/purpose), working with that group to safely achieve the goals of the gathering and to ensure open communication is maintained so that everyone is aware of the plan (Drury, 2015).

Planning and communications

It is important to understand that any crowd arriving at a venue will be influenced by the design and information provided upon approach and, from an organisers point of view, that these two elements can be controlled (Still, 2016).

Organisers can ensure that ingress and egress routes are of sufficient of a size and condition that they will safely accommodate the anticipated crowds. Furthermore, alterations to the design can be made in advance so as to minimise/eliminate any congestion that may occur.

After the organisers have established an overall plan, the information must be clearly packaged and communicated to the public in a manner that prevents any ambiguity. This includes media messaging in advance of the event and installing prominently displayed signage on site that directs crowd movement.

Not only does good communication provide people with information (facilities, locations, correct exits in an emergency), it can also enhance the relationship between organisers and attendees, in a virtuous relationship of trust (Drury, 2015).

Furthermore, a plan to address emergencies should be established in advance of the event. This plan must be communicated to the support staff so that they are informed on what is specifically expected of everyone. It has been proven that there are significant improvements in evacuation rates when trained people are in the crowds providing proper communications (Pelechano, O'Brien, Silverman & Badler, Norman, 2005).

Tools

Members of the Ottawa Police Service (OPS) Event Planning Section have always developed detailed plans in advance of major events. Those plans always address crowd management issues, but they lack the necessary scientific foundation. In May 2016 OPS hosted Dr Keith Still, an expert and professor in crowd science. The tools acquired during the training identified that scientific support was lacking in the Canadian process.

The *DIM-ICE Model* designed by Dr Still commits planners to addressing the three key categories related to crowded space (design, information and management – DIM). These categories are separately addressed in the model during the various phases of an event (ingress, circulation, and egress – ICE). This model can, and should, be developed in advance of an event for both normal and emergency situations (since both are dynamically different and, in many cases, differ significantly).

A *RAMP Analysis* (Routes, Area, Movement, Profile) is used to identify event characteristics that includes, among other elements, potential high-density areas, capacity restrictions and crowd types. This analysis highlights areas of concern so that the planners/organisers can properly address identified issues.

Another important tool is *Decision Support Analysis*. This tool is used to determine high and low risk situations for any number of comparable elements. In relation to crowds, it can be especially useful in highlighting high risk areas in terms of flow rates and crowd density during different periods of time. Using this tool will allow a planner to effectively adjust security and support to those areas for a given period of time.

Conclusion

Safety and security are paramount during any event. To achieve this, appropriate planning and organization is essential to ensure any potential risks are minimised and manageable. Since organisers are generally focused on attracting more patrons, crowd management must be considered at every stage of the planning process. Properly forecasting potential crowd issues through a comprehensive pre-event analysis will ensure that the event runs smoothly, the patrons enjoy a positive experience, and the organiser achieves the successes intended.

REFERENCES

Agnes, O. Minu 2012, Proxemics: The Study of Space, The Indian Review of World Literature in English, Vol. 8, No 1, pp 1–7. Available from http://worldlitonline.net/proxemics-the-o.pdf [worldlitonline.net] (09 August 2016).

Alnubulsi, Hani & Drury, John 2014, 'Social identification moderates the effect of crowd and Boundaries", Man's image in Medicine and Anthropology.

Density on safety at the Hajj', Proceedings of the National Academy of Sciences of the United States of America, vol. 111 no. 25, pp. 9091–9096 Available from http://www.pnas.org/content/111/25/9091.long [pnas.org] (09 August 2016).

Drury, John, 2015, Bringing Crowd Psychology into Event Safety Management. Available from http://www.festivalinsights.com/2015/07/bringing-crowd-psychology-event-safety-management [festivalinsights.com] (09 August, 2016).

Government of Canada, Statistics Canada 2016, Civil court survey, general civil cases by level of court and type of action. Available from: http://www5.statcan.gc.ca/cansim/a26?lang=eng&id=2590013 [statcan.gc.ca] (09AUG16).

Hall Edward T (1989), *Beyond the Culture*, Anchor Books, New York.

Hall Edward T (1963), *Proxemics: The Study of Man's Spatial Relations*, http://www.uni-kiel.de/psychologie/ispp/doc_upload/Reicher_crowd%20dynamics.pdf [uni-kiel.de] (09 August, 2016). International University Press, New York.

Le Bon, Gustave, (1896), The Crowd – A Study of the Popular Mind. Available from https://socserv2.socsci.mcmaster.ca/~econ/ugcm/3ll3/lebon/Crowds.pdf [socserv2.socsci.mcmaster.ca] (09 August, 2016).

Martin's Annual Criminal Code -2016 Police Edition, 2010, Thomas Reuters, Aroura.

Pelechano, Nuria & O'Brien, Kevin & Silverman, Barry & Badler, Norman 2005, Crowd Simulation Incorporating Agent Psychological Models, Roles and Communication, University of Pennsylvania Scholarly Commons. Available from http://repository.upenn.edu/cgi/viewcontent.cgi?article=1033&context=hms [repository.upenn.edu] (09 August, 2016).

R v Sharp (1984).

Reicher, Stephen n.d., The Psychology of Crowd Dynamics, School of Psychology, University of St Andrews. Available from https://pdfs.semanticscholar.org/2ed1/dacbd7177d6c1709dda3a8254e7e58a6d1af.pdf [pdfs.semanticscholar.org].

Still, G K (Dr.) 2016, Book – Introduction to Crowd Science. Available from: http://www.gkstill.com/Support/Book.html [gkstill.com] (15JUL2016).

Still, G K (Dr.) 2016, Crowd Modelling: Crowd Safety in the Complex and Built Environment. Available from: http://www.gkstill.com/Support/Links/Documents/dim---ice.pdf [gkstill.com] (15JUL2016).

Still, G K (Dr.) 2016, DIM-ICE Risk Analysis. Available from: http://www.gkstill.com/Support/WhyModel/dimice.html [gkstill.com] (15JUL2016).

Still, G K (Dr.) 2016, PhD Chapter 1- Introduction. Available from: http://www.gkstill.com/CV/PhD/Chapter1.html#previous-photo [gkstill.com] (15JUL2016).

Still, G K (Dr.) 2016, PhD Chapter 3 – Crowd Dynamics. Available from: http://www.gkstill.com/CV/PhD/Chapter3.html [gkstill.com] (15JUL2016).

Still, G K. (2013). Introduction to Crowd Science. Boca Raton, FL: CRC Press.

Chapter 15

Example case study

Knowledge without application is like a book that is never read.

Christopher Crawford

In this section, we show one example case study from a practitioner, using the modelling tools for major events. We have reproduced the work in greyscale.

On the website (https://www.routledge.com/Applied-Crowd-Science/Still/p/book/9781138626560) you will find the other cases studies and the delegates original work (in full colour).

When presenting your work, we recommend the following procedure.

1. Select a relevant map, plan or diagram
2. Import to your word processor, presentation tool or spreadsheet
3. Select image and then reduce the saturation to zero (this makes it greyscale)
4. Then add your Red/Orange/Green lines and areas

That will make your graphics easier to distinguish from the background.

Due to the print limitations, we can only reproduce the colour images in grey scale. Full colour versions are available in the online version.

KRISTIN FINN – RUGBY WORLD CUP, FAN ZONE

An evaluation of the Rugby 2015 World Cup Fan Zone site, Exeter

> Kristin Finn is a Lecturer in Event Management for the Department of Tourism and Hospitality at Plymouth University, and her main teaching roles are in crowd management, research and consultancy, and operational planning for events. Specifically, her research interests lie in matters relating to crowd behaviour, managing safety and the relationship between crowds and audience experience at events.

DOI: 10.1201/9781351053068-15

Prior to joining Higher Education, she worked in consultancy for almost ten years, most notably developing project evaluations, monitoring frameworks and organisational strategies for local government as well as visitor and audience behaviour studies for tourism, cultural and events organisations in the public and private sectors. This range of experience informs her teaching contribution and areas of professional interest.

Introduction

As a host city for the 2015 Rugby World Cup in September, Exeter City Council announced the provision of a free 5,000 capacity fan zone site located in Northernhay Gardens. This was made available for the thirteen match days over the six weeks of the event in the heart of the city centre for fans and all those who were unable to secure tickets (Heart of Devon, 2015). Northernhay Gardens is a popular location for outdoor events in the city owing to its central location and proximity to a variety of transport links. The Gardens are utilised annually for the Exeter Festival of Southwest Food and Drink (Exeter Food and Drink Festival, 2015) as well as Big Screen in the Park (Exeter Phoenix, 2013) amongst other events. The site accommodates up to 5,000 people, with facilities which include a big screen, food and beverage outlets, merchandise stalls, toilets and information facilities (Hatfield, 2015). However, the site is not without its challenges. This paper analyses the suitability of the Fan Zone site, exploring potential issues associated with crowd flow, congestion and site design. This is then utilised to map risks and present recommendations for an appropriate crowd management strategy for the event.

Analysis of Rugby Fan Zone site, Exeter

Tum et al. (2006) suggest that the three fundamental on-site issues to be addressed through layout planning are to optimise movement, reduce congestion and maximise the use of space in an event site. Underpinning this concept is the importance of site planning in ensuring crowd safety and positive customer experiences at events. Rutherford-Silvers (2008:247) argued that the design of the event setting must strive to encourage desired behaviour in a range of ways linked to the functional environment of the event.

Building on this, Still (2014) identified the need to evaluate several aspects in relation to crowd dynamics. First is RAMP analysis, which looks at evaluating event routes, areas, movement, and profiles. Second, is modelling crowd risk according to ingress, circulation and egress as well as design, information and management (DIM-ICE). Third is decision-support analysis. Fourth is the need to develop risk maps for the event site. This analysis has been conducted for the Fan Zone site.

RAMP analysis[1] – Routes (directions)

Crowds will ingress and egress the Northernhay Gardens site from three main access points (Figure 12.4) the High Street gate (A), the train station gate (B) and three smaller entrances into the walled park from the Exeter Phoenix Arts Centre, the library and the castle (adjacent to Northernhay Gardens). These converge into one gated entrance (C) taking the attendees to the main Fan Zone site. Two main roads wrap the entrances to the site, Queen Street and the High Street, off which there are three multi-storey car parks, a train station, a bus station and seven bus stops. Each of these transport points is within a two-minute walk of one of the gates to the venue.

Areas

The area around each ingress and egress gate will be manned by stewards to monitor visitor numbers and will include an information point. For this reason, it is expected that these areas will experience casual, walking and ambulatory crowds (Berlonghi, 1995). The challenge for organisers will be to facilitate a steady flow through these nodes. The large screen integral to the fan zone is positioned at the west end of the park, close to Gate B (Queen Street Entrance) with restricted staff-only access directly behind it. A first aid point is also positioned by Gate B. In front of the screen and stage in the centre of the fan zone is expected to be a free standing and seating area (GA) with toilet facilities to the north boundary of the site. Crowds in this area may be seated, standing, migrating or queuing, and at this point, the crowd will most likely become dense. The area nearest to Gate A (High Street entrance) will house the food and beverage concessions and further toilet facilities. It is expected that queues may form in this area, with crowds becoming densest at peak arrival times and between scheduled entertainment slots or matches. A detailed map of the site areas is shown in Figure 15.1.

Movement

Flow rates will be wholly dependent on the timing of the matches to be shown on the big screen and the entertainment available in the fan zone well as the site opening time of the site each day. Unfortunately, without full access to this information, flow rates can only be crudely estimated. However, a fan zone schedule for Tuesday 29th September 2015 (Heart of Devon, 2015) was available and has been used to estimate flow rates. Depending on match times and entertainment available, flow rates are likely to also be influenced by public transport arrivals from the arterial routes (trains and buses/park

[1] This is a case study event only – access to specific site dimensions was unavailable to inform maps.

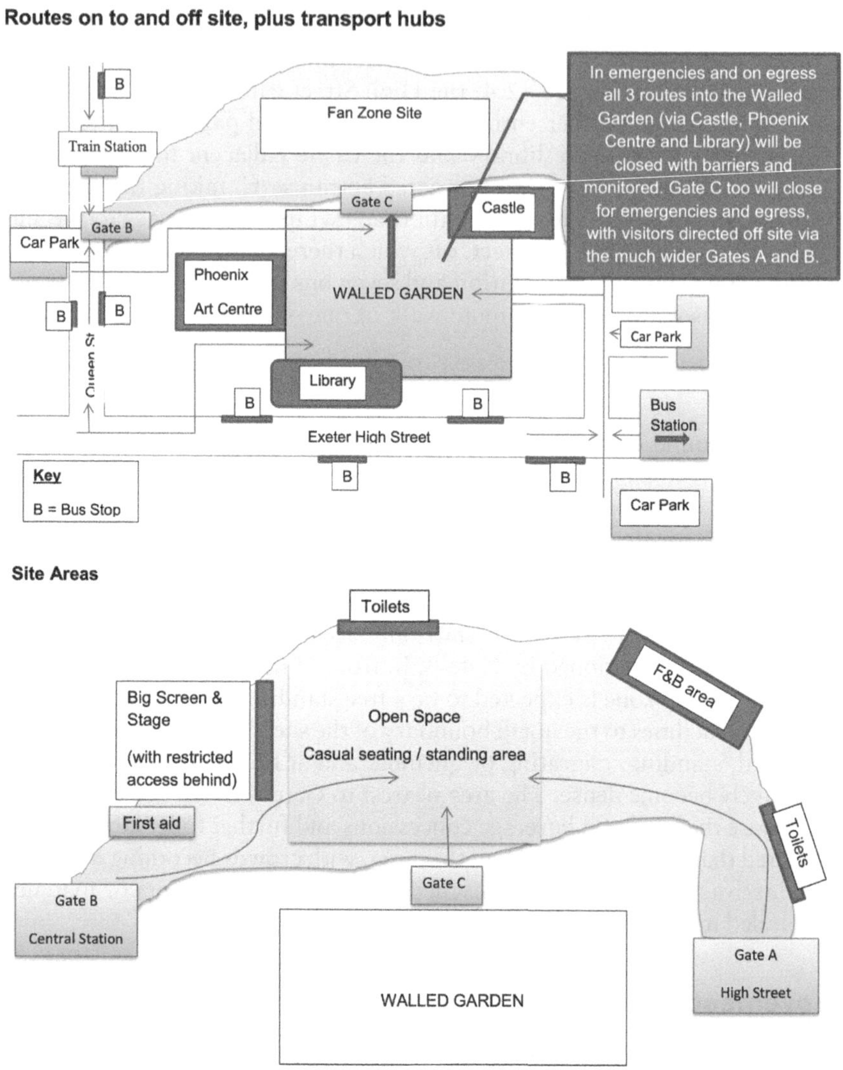

Figure 15.1 Site, routes, hubs and areas.

and rides). The stage programme and public transport[2] options for the day to help estimate flow rates (Figure 15.1 (top)). The fans will be pulsed from the arrival sites towards the fan zone in time with schedule enabling some intelligence to be gained from bus and train timetables.

Fan zone stage programme – September 29th (Heart of Devon, 2015)

[2] National Rail Enquiries 2015; Stagecoach Exeter 2015

11:00–13:00 – Schools music line up
16.00–16.10 - Fan zone Advantage new artists and bands
16:15–16:45 – Live Screening Rugby World Cup 2015
16.45–18.45 – Tonga v Namibia live screening from Sandy Park, Exeter
18.45–19.00 – Phoenix Film
19.00–20.30 – Zuri Aura-5-piece Afrobeat/ Reggae/Dub band Zuri Aura
Train Arrivals Bus Arrivals
18, 19, 28, 37, 51 and 52 16 routes arriving every 10–15 minutes
Minutes past every hour

As there is little entertainment taking place ahead of the match on the specified day, visitor flow to the site until lunchtime is expected to be relatively low, perhaps centred mostly on walk-ins. A slow-to-steady flow at the gates and around the site is likely from opening until approximately two hours before the match starts and whilst the match is being played, when it is anticipated that most visitors will have arrived.

Increased flow is likely to be experienced at the gated entrances in the 2 hours leading up to match airtime owing to public transport arrivals, with 6 trains arriving per hour and 16 bus routes arriving in the centre every 10–15 minutes per route. At these times the standing and seating area will begin to fill. However, arrivals by car and walking will provide the largest proportion of visitors, with walkers being local to the area and high car use identified as a key characteristic of Exeter's day trip market (Heart of Devon, 2006). Car users (including park and ride visitors) and walkers are expected to greatly increase in numbers during this 2-hour time frame with peak flow likely to occur within one to one-and-a-half hours of the match start. At this point it is expected that the site will be near to or full, reaching its 5,000 people capacity. It is likely that egress bottlenecks will also be experienced at the match end and following the evening band entertainment post-match.

Profile

Based on previous events held in this venue including the Big Screen in the Park (Exeter Phoenix, 2013) and based on the common typologies of sports fans (Bladen et al., 2012), the event is likely to attract a range of ages and group types, including families, friends, and sports fans of varying degrees of commitment. Arguably, these crowds will mostly be revellers and expressive, enjoying the atmosphere (Berlonghi, 1995; Rutherford-Silvers, 2008) or as Canetti (1973) would define them, a "feast" crowd.

It is important to also understand the fan dynamic and the sports event subculture which can indicate potential for rivalry between supporters and excessive consumption of alcohol that can exacerbate certain situations (Dun, 2014; Green and Chalip, 1998; Henderson, n.d.). With the large range of audience types likely to be in attendance, a strong steward and security presence as well as emergency strategies involving the police and other emergency services is required. However, it can also be noted that the preponderance of rugby fans will provide a less exuberant and violent type of spectator than that of a football crowd.

DIM-ICE evaluation

The design, information and management of the Exeter Rugby World Cup Fan Zone event have been analysed according to ingress, circulation and egress scenarios for both normal and emergency periods. As can be seen in Tables 15.1 and 15.2, the event site is suitable for its size and could be considered as relatively low risk overall, with appropriate procedures in place generally for ingress, circulation and egress in both normal and emergency scenarios. The main areas for concern and high risk are linked to the restricted width of Gate C and the subsequent need to close and monitor this gate for egress. Moreover, the Walled Garden beyond Gate C is dimly lit at night which could present a risk to visitors. This demonstrates the need for its closure (from all access points) as well as the Gate C exit directly attached to it.

Congestion/risk maps for the site

According to Helbing and Mukerji (2012), crowd surges and safety scenarios are not always avoidable as crowds are often unaware of the domino-effect dangers involved in dense crowd situations. Following the RAMP and DIM-ICE analysis conducted for the Rugby World Cup Fan Zone site, congestion risk maps have been produced below to identify low (green), medium (amber) and high risk (red) congestion areas at different points in time within the event. Figure 15.2 considers the changing crowd scenarios linked to the event opening time (11 am), peak ingress (2–3.30 pm), match/entertainment time (4–8 pm) and event closure/egress (8.30 pm) (Figures 15.3 and 15.4).

Decision support matrix/analysis

From the decision-support analysis conducted, high risks of crowd density (Table 15.3) are centred around the peak ingress and egress times on the gates and during match showing and entertainment for on-site areas. Lines of communication are expected to be good generally (Table 15.4) although ensuring clear separation of tasks is crucial and audience clashes and acceptance issues may be observed owing to rivalries and intoxication. Direct liaison with the police will require swiftly action by organisers as well as issues are raised by staff team members.

Table 15.1 DIM-ICE analysis – normal

Normal	*Ingress*	*Circulation*	*Egress*
Design	Free entry non-ticketed event whereby gate staff count visitors onto site. Wristband entry system to help ensure capacity is not breached. Visitor numbers to be monitored upon entry. Multiple (3) ingress / egress routes. Signed transitory routes to view area in central open space	Flows expected to build up to start of the match. Site map leaflets provided on entry. Signage position in sight line of entry and at key places to move visitors through the main event space. Barrier lane entry to be deployed primarily to accommodate the peak flow time 2 hours prior to match start Stewards to monitor bottlenecks at entry and near F&B and toilet facilities. F&B and toilet facilities likely to become congestion directly before match start, during half time, and at match end.	Bottlenecks likely at end of match for the three key egress points (Gates A, B and C). Sufficient width to accommodate flow at Gates A and B. Gate C is narrow so closed for egress. Flow redirected by stewards to either Gate A or B.
Information	Signs to open viewing area and stage. Information booths located next to entry gates. Pre-event publicity to provide key transport information, directions and expected crowds with a capacity limit of 5000 maximum	Signage aimed at directing people around the site. Stewards deployed to monitor busy areas on site	Signage routes for egress off-site through Gates A and B, and away from the venue to key transport hubs clearly marked. Staff / stewards clearly identifiable to move visitors off-site. Site closing information shared on the Big Screen
Management	Stewards / security to count numbers entering site and distribute wristbands Stewards present in information booths. Stewards present in viewing area. Monitoring of arrival flows to site in city centre by organisers and police	Good communication systems in place Two-way radio communications between gate staff organisers, plus organiser contact with emergency services (when required). Clear routes and directions for visitors to and around the site Strategy for crowd congestion at key times in key locations in place	Good CCTV around the City Centre to monitor egress flow Communication between organisers and police Fully briefed staff and volunteers to aid smooth egress.

Table 15.2 DIM-ICE analysis – emergency

Emergency	*Ingress*	*Circulation*	*Egress*
Design	Pinch point at Gate C from Walled Garden entrance to site at peak times potentially. All other gates easily accessible with sufficient space for visitor numbers expected. Ample off-street space to accommodate barrier queues on entry at all three gates.	Hazards and restricted areas clearly marked (particularly restricted area behind the Big Screen and stage). Sufficient venue space for capacity of 5,000. Stewards deployed to manage congested areas at peak times.	Bottlenecks and congestion likely if visitors try to exit through Gate C – therefore gate to be closed for egress at end of event with staff deployed to gate to redirect flow in opposite directions for Gates A or B. Walled Garden dimly lit at night, so all entrances closed off and monitored to avoid visitor injury.
Information	Good signage to venue and good communication channels in place.	Big screen used for messages to crowd and PA system in place.	Staff to relay information updates via Big Screen and Tannoy. Police presence outside to help manage emergency egress off-site in areas around High St. and Queen St.
Management	Staffing levels are good due to a strong volunteer base as well as core event staff. Careful monitoring of entrance gates and city centre approaches. Strong communication with police.	Careful monitoring of crowd congestion. Clear roles defined in a circulation-based emergency for staff working in specific event zones. Event stop procedure in place if required.	Good CCTV around city centre to pinpoint emergency scenarios. Clear. Stewards to manage safe and swift egress on-site until venue is cleared.

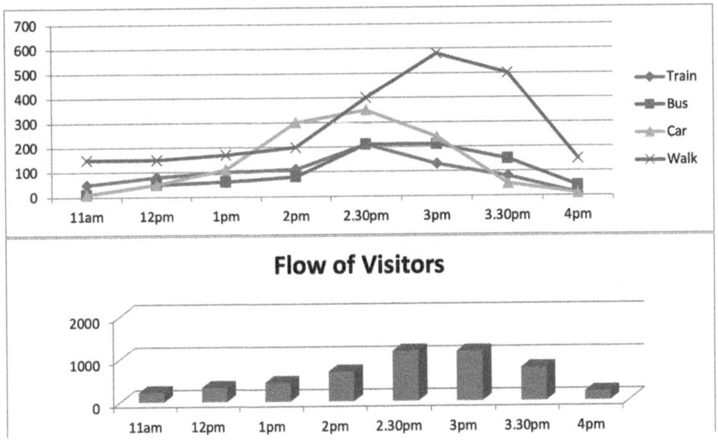

Figure 15.2 Anticipated arrival flow (modal split) and likely visitor flow.

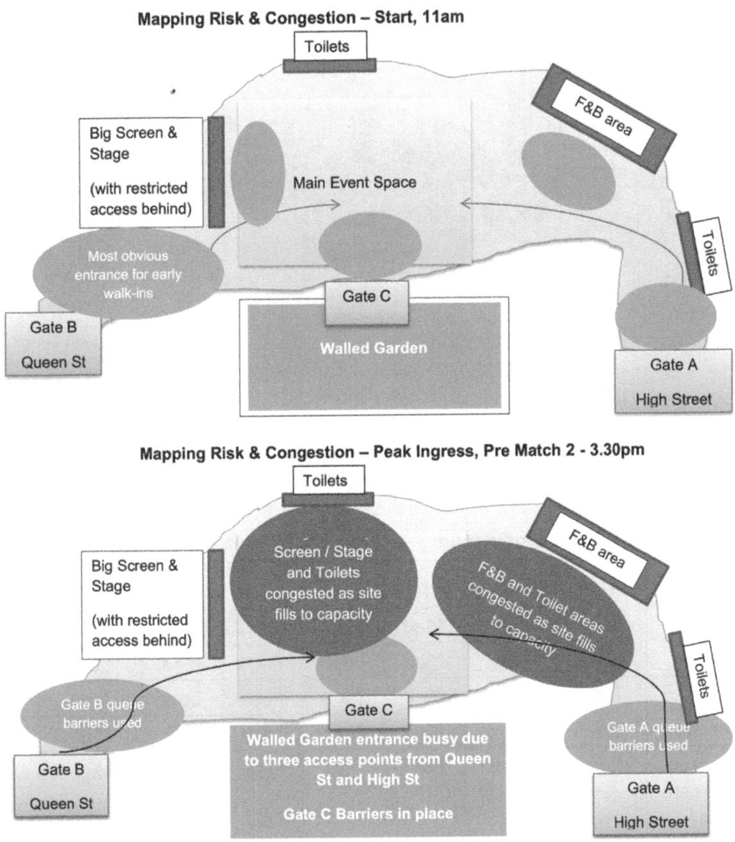

Figure 15.3 Mapping risk and congestion (start time 11 am and pre-match 2–3:30 pm).

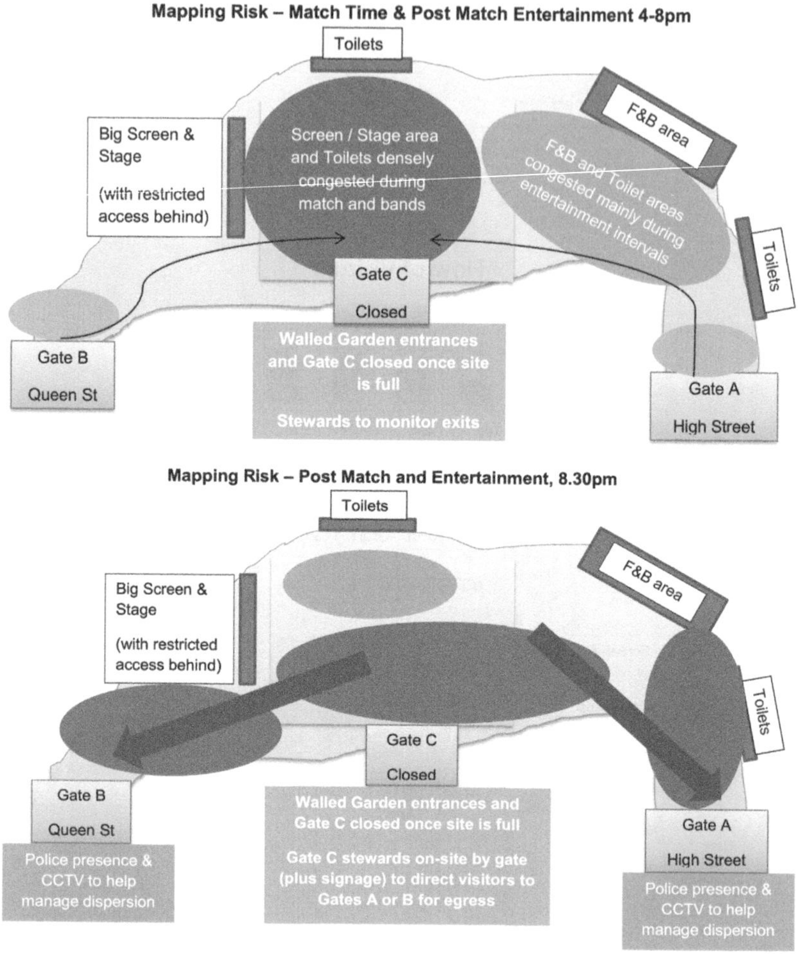

Figure 15.4 Mapping risk and congestion (4–8 pm and post-match 8:30 pm).

Table 15.3 Crowd density/movement – risk profile by area

	11 am	*12 pm*	*1 pm*	*2 pm*	*3 pm*	*4 pm*	*5–8 pm*	*8.30 pm*
Gate A	Med	Med	Med	High	High	Med	Low	High
Gate B	Low	Low	Med	High	High	Med	Low	High
Gate C	Low	Low	Med	Med	High	Low	Low	Low
Pathways	Low	Low	Med	Med	High	Med	Med	High
Main stage and screen area	Low	Low	Med	Med	High	High	High	Med
Toilets	Low	Low	Low	Med	Med	High	High	Med
F&B area	Low	Low	Med	Med	High	High	High	Low

Table 15.4 Communication decision-support matrix

	Organiser	*Police*	*First Aid*	*Staff*	*Security*	*Audience*
Organiser	Radio sets Clear comms, clear emergency comms plan	Liaison in control room	Clear briefing & incident plan, radios	Clear tasks & line of comms, radios	Clear tasks & line of comms, radios	Big Screen and gate info, loudspeakers, Clear online info
Police	Clear terminology, radios, CCTV	Clear tasks and structure	Clear separation of tasks	Separation of tasks, need for co-working	Separation of tasks, need for co-working	Potential issues with fans or those intoxicated
First Aid	Comms for incidents, radios, separation of tasks	Liaison if needed through organisers	Clear roles and responsibilities, radios	Clear separation of tasks	Clear separation of tasks	High acceptance
Staff	Clear tasks and line of comms, radios	Liaison through organisers	Clear separation of tasks	Clear roles and responsibilities, radios	Clear separation of tasks, radios	potential issues with those who are intoxicated
Security	Clear tasks, radios, clear line of comms	Radios, liaison in control room	Radios, clear separation of tasks	Radios, clear separation of tasks	Clear roles and responsibilities, radios	Potential issues with those who are intoxicated
Audience	Personal contact, big screen, online	Medium acceptance	High acceptance	Medium acceptance	Potential issues with those who are intoxicated	Potential clashes – rivalry and intoxication

Conclusion and recommendations

The analysis presented within this paper sets out the required crowd management and control planning that should be implemented in both a normal and emergency crowd situation for the 2015 Rugby World Cup Fan Zone event in Exeter using the tools identified. The site of Northernhay Gardens is event-appropriate owing to its central location with close links to key transport hubs and several routes onto site to alleviate the pressure on a single gate. In the case of an emergency, lines of communication, roles and tasks must remain clear amongst all event stakeholders and in the case of an emergency egress, organisers and staff should act to closely and monitor Gate C as well as the three ingress points to the Walled Garden, redirecting visitors to leave the site through either Gate A or B instead.

Peak flow is expected within the 2 hours leading up the start of the match on the gates as well as upon egress. Peak flow is also expected in the main event space in front of the screen and stage and by F&B and toilet facilities during the match and also during post-match entertainment.

This event will attract a broad range of visitors, and whilst most will be peaceful, it is anticipated that clashes could occur among fans or those who are intoxicated presenting issues in terms of managing audience communication. At egress in particular, a greater police presence may be required to safely monitor, manage, and escort fans from the premises.

REFERENCES

Berlonghi, A.E. (1995). Understanding and planning for different spectator crowds. *Safety Science*. **18**. 239–247.

Bladen, C., Kennell, J., Abson, E. & Wilde, N. (2012). *Events Management: An Introduction*. Oxon: Routledge.

Canetti, E. (1973). *Crowds and Power*. London: Penguin Books. Translated by Carol Stewart.

Dun, S. (2014). No beer, no way! Football fan identity enactment won't mix with Muslim beliefs in Qatar 2022 World Cup. *Journal of Policy Research in Tourism, Leisure and Events*. **6** (2). 186–199.

Exeter Food and Drink Festival (2015). [online] Available at: www.exeterfoodanddrinkfestival.co.uk Accessed on: 19.06.15.

Exeter Phoenix. (2013). Big Screen in the Park. [online] Available at: www.exeterphoenix.org.uk/big-screen-in-the-park/ Accessed on: 20.06.15.

Green, C. & Chalip, L. (1998). Sport tourism as the celebration of subculture. *Annals of Tourism Research*. **25** (2). 275–291.

Hatfield, V. (2015). Exeter City Council Tourism Officer. Personal Communication.

Heart of Devon (2006). Exeter Visitor Strategy 2007–2010. [Online] http://www.heartofdevon.com/xsdbimgs/EXETER_VISITOR_STRATEGY_2007-10[1].pdf Accessed on: 29.06.15.

Heart of Devon (2015). Exeter Fanzone. [Online] Available at: http://www.heartofdevon.com/exeter/exeter-fanzone Accessed on: 21.06.15.

Helbing, D. & Mukerji, P. (2012). Crowd disasters as systematic failures: analysis of the Love Parade disaster. *EPJ Data Science* **1**, (7). 1–40.
Henderson (n.d.). Fan Behaviour – A World of Difference. [Online] Available at: http://pc.parnu.ee/~htooman/EuroChrie/Welcome%20to%20EuroCHRIE%20Leeds%202007/EuroCHRIE%20Leeds%20Conference%20Preceedings/Full%20Papers/Henderson, %20Stephen.pdf.
National Rail Enquiries (2015). Live Departures. [Online] http://ojp.nationalrail.co.uk/service/ldbboard/arr/EXC Accessed: 25.06.15.
Rutherford-Silvers, J. (2008). *Risk Management for Meetings and Events*. Oxon: Routledge.
Stagecoach Exeter (2015). Exeter Network Route Map. [Online] Available at: https://www.stagecoachbus.com/uploads/exetermap.pdf Accessed: 25.06.15.
Tum, J., Norton, P. & Nevan Wright, J. (2006). *Management of Event Operations*. Oxford: Butterworth-Heinemann.

Chapter 16

Crowd science and Covid 19

> Imagine the action of a vaccine not just in terms of how it affects a single body, but also in terms of how it affects the collective body of a community.
>
> Eula Biss

We wrote the first draft of this book before the COVID-19 pandemic and the international lockdowns. For obvious reasons, we put this publication on hold during the pandemic, as the world was changing, and events has shut down, there was uncertainty and revisions of crowd safety, such as social distancing, masks and capacity calculations. At the time of submitting this version to the publisher, the pandemic has changed the nature of events, but there were signs that "normal" may return.

During the initial stages of the pandemic, we developed a course for the members of the Institute of Place Management[1]. This was based on modelling data for contagious diseases and crowd infections and used during modelling the information spread in various financial markets. We had a level of understanding of both the SIRS model, and the impact on crowded spaces. We also had the relevant experience of creating, hosting, and supporting eLearning programme. This was an accredited course, developed for the Institute of Place management members only.

While writing this section, we are still in an uncertain stage of managing crowds following the Covid Pandemic.

DR. SYAN SCHAAP – COVID-19, CROWD MANAGEMENT

About Syan

Syan is one of the leading experts in The Netherlands on event and crowd safety. He holds a master's degree in Public Administration and has over 15 years of experience as a consultant, advising local

[1] https://www.placemanagement.org

DOI: 10.1201/9781351053068-16

authorities, event organisers and location owners. Syan excels in the evaluation and revision of safety planning for large scale public events. He's also involved in the safety planning in crowded city centres and touristic locations. Syan combines strategic advice with a practical, operational view on the planning and execution. He developed the RISKOM method for the multidisciplinary risk assessment.

As a trainer, Syan has delivered many courses on crowd science, event safety, fire safety, public order management and risk assessment. Since 2016, he works together with Prof. Dr. G. Keith Still in delivering the level 5 course in The Netherlands and Belgium. He teaches this course in Dutch under supervision of prof. Still Syan also teaches in the minor Crowd Safety at Breda University of Applied Science (BUAS), a unique minor program developed by prof. Still, ESI and BUAS. He develops and delivers the courses on event safety advising for the national police and Safety Regions in The Netherlands. Syan is director and co-owner of the Event Safety Institute.

The Event Safety Institute (ESI) is an independent centre of expertise with a strong focus on event safety. ESI has been founded in the Netherlands in the year 2014 and has a team of 12 experienced consultants. At ESI our standard approach is to apply the scientific knowledge and analysis techniques from Prof. Dr. G. Keith Still primarily to the safety of events, either through education of the relevant stakeholders or through our direct consultation work. The Covid pandemic proved these techniques are very useful for other sectors since all societies became a crowd management challenge.

Covid-19 pandemic: society as a crowd management challenge - strict rules

Shortly after the outbreak of the Covid-19 pandemic in March 2020, the Dutch society was confronted with the strict government measures designed to mitigate the effects of the pandemic which included for example a total ban on events. This also severely impacted our operations in eliminating 90% of our work stock in a very short period. Such operations included major events such as the first Formula 1 race in the Netherlands since 1986 for which 104,000 tickets had already been sold and UEFA Euro 2020. At the same time, we were also contributing towards other iconic events including the national King's Day celebration in Maastricht, the opening week of the Eurovision Song Contest in Rotterdam and several dance festivals with an international appeal for which our country is well known for. The events were all banned overnight and the rest of society at the same time was confronted with other strict rules predominantly based upon the concept of social distancing. This concept consists of measures such as keeping 1.5 m towards each other, a ban on group formation as well as a standard of 10 m^2

per visitor in shops. As the pandemic persisted this concept was also dubbed as the "new normal", a paradigm that also brought new opportunities for ESI: The entire society had become a crowd management challenge because of this "new normal" on factors such as capacity and density of crowds.

Aid to cities

Due to the rules and regulations management of crowds was immediately warranted at the inner-city shopping areas. After building our expertise in aiding these areas towards a safe operation, we offered our help to various cities in the Netherlands. Based on our in-depth knowledge of event and crowd management combined with our new experiences of managing Covid-19 challenges, our offer was gratefully accepted. The next period for ESI thus consisted of executing mid- to large scale projects in Eindhoven, 's-Hertogenbosch, Alkmaar and Maastricht. Also, smaller scoped projects where successfully completed for many other cities and municipalities such as Amsterdam (Red Light District) and a multitude of other locations with a broad scope from regulating parks to the opening of a natural ice rink in the city of Breda.

What could we contribute to these cities? Using the insights of Prof. Dr. Still, we created a clear overview of the new situation: the limitations the Covid-19 measures induced in terms of visitor capacity, the consequences social distancing for the layout of the outdoor space and the need to prevent conflicts in the use of available space. For example, enlarging a terrace to enable social distancing quickly results in less available space for the passing crowd flow. Such could lead to potentially dangerous situations when the pedestrian flow would be forced to walk on busy roads with motorised traffic. We assisted cities to identify such bottlenecks, provided solutions and contributed to operationalizing the efforts to build an efficient but most of all safe crowd system.

Directed approach

Various municipalities set up joint project teams in which departments such as Public Safety, Traffic, City Marketing, City Management and Communication drafted the measures designed to mitigate the effects of Covid-19. The directed approach of ESI ensured that they, as a team, were properly informed about the consequences of such measures towards managing the crowd.

As an example, it was generally assumed that the use of city hosts could assist the public in following the social distancing rules and thus many cities put such hosts on the street. We informed the joint project teams, based upon our expertise and experience, that such city hosts should be supported with clear signage that stand out and appeal to visitors to be effective. They must also be backed up through bylaw enforcers or the police. Even with a solid crowd management plan it was found that certain areas remained too busy to enable social distancing. Towards such overcrowded locations we provided clear calculations regarding crowd capacity and -flow, in some cases supported with crowd simulations constructed with the software of InControl Simulations.

Phased approach

ESI's approach was based on our vision that the first phase of interventions should be deployed quickly, through people and resources often associated with events. Such as the deployment of event security guards, matrix signs, fences to close narrow alleys, and by using stripes and arrows on the ground to designate directions. We designated such measures as "the event phase". This labour-intensive (and thus expensive) event phase should have evolved over time towards a more economically maintainable phase. In this so called "consolidation phase" the labour-intensive interventions could then be scaled down as technology would be introduced to continuously monitor the activity in an area. Such data would enable city workers to draft long-term policies regarding issues such as car traffic in the city centre, about cycling routes and the need for redesigning and increasing recreation space.

However, it was found that the dynamics of the Covid pandemic made it very difficult for municipalities to plan with a scope of more than a few weeks. Thus, at present time, more than a year into the crisis, the phase of consolidation is not reached by many cities. Contributing to this effect is the practice of the national government to draft nationwide mandatory measures that then must be implemented by the local authorities. These measures changed from month to month or even weeks as they were predominantly based upon the volatile number of Covid infections. A sudden drop in the number of infections shortly before the summer of 2020 resulted in the effect that many cities (that finally had reached a stable event phase) where confronted with a strong call for abruptly reducing these measures. This in combination with fatigue among the project team members that had already been intensively involved with the approach, led to a strong delay in long-term planning. Thus, the investments in crowd monitoring within the city centre have been postponed and most other measures are still regarded as temporarily instead of being part of a permanent crowd monitoring system.

DIM-ICE

For many clients ESI has satisfactorily implemented the DIM-ICE risk analysis. We introduced the concept that within every busy place or location, there are always three factors that can be adjusted: Design, Information and Management. Using a solid risk analysis and the DIM-ICE risk matrix this enabled determining the most effective approach.

Within the city centre the factor Design is primarily about designating the available space of the shopping and leisure locations based upon the social distancing concept which severely limits the capacity of such areas. This can lead to restrictions on terraces, parking spaces and pavement signs outside of the shops. Design has been found to be a balancing act by the municipality workers in deciding which use should take precedent over other uses, e.g., routes vs waiting areas or parking spaces vs terraces.

During this pandemic the factor Information is in a large part aimed at public communication and at nudging towards certain desirable behaviour. A good example of this nudging is the marking of queue or bicycle parking space on the streets which was found as a low-cost intervention with good results especially when combined with the deployment of staff. General public communication in the Netherlands increasingly focused on compliance towards the basic measures: washing, social distancing, staying at home (later: testing). As the pandemic persists the focus increasingly is aimed at behavioural support interventions.

The factor Management of the hustle and bustle in the inner cities has several facets. At street level, it consisted of the city hosts (and sometimes also actors) who are used to remind visitors of the rules and to present information about alternative routes. At city level, it is about keeping an overview of the influx of visitors, the crowds relative to the limited capacity of the areas. Management also intervened within those crowds when there is a potential overcrowding. It was found that many cities struggle to find interventions that are effective and acceptable to the visitors of such locations so ESI has drafted many crowd management plans to assist those cities in their efforts.

Managing visitors in recreational areas

As restaurants, cafes, museums, shops and gyms all had to close in the autumn of 2020, new challenges arose due to the limited availability of outdoor recreation. Many urban parks thus became overcrowded, which ultimately resulted in regular interventions by the riot police to evict the visitors of these locations. In our analysis this repressive (and aggressive) form of crowd control was performed in places where it was foreseeable to occur. ESI wrote blogs titled "don't sweep but regulate" and about "The Park as an event", to encourage local authorities to approach the parks as festival grounds and to apply interventions that are like those at free-access pop music festivals.

Somewhat similar problems arose within the countryside where the amount of available parking space at national parks was found to be a bottleneck. Long traffic jams resulted in frequent calls to not travel to these locations while at the same time the locations itself were not overcrowded. The increased appeal of these parks also resulted in many visitors leaving the official paths and littering. At current time many interventions are therefore directed towards proper visitor management of these areas.

National guidance

As a centre of expertise, ESI was also tasked with contributing towards two national guidelines regarding crowd management. In June 2020, ESI issued the guide "Regulating crowds in public space" for the Association of Dutch Municipalities (VNG). This guide enabled us to introduce our concepts based upon the factors of Design, Information and Management to a nationwide audience. We also included means towards differentiating communication towards visitor profiles, such as city residents, regional visitors and tourists from abroad. We provided blueprints for the layout of streets including queue areas and specific attention to the special needs of visitors with a physical disability.

The second guide we contributed to was aimed at the tourism sector. The Netherlands Bureau for Tourism and Conventions (NBTC) had recently focussed their efforts on the balance between the concepts "Experience", "Quality of Life" and "Safety". Thus, even before the covid pandemic, visitor management was becoming an important new theme for this NBTC. Within the Visitor Management Guide published by the NBTC, ESI has outlined the approach based of the DIM-ICE risk analysis.

The guide contains examples of a DIM-ICE risk analysis for a tourist city and a popular recreation area. At present time NBTC continues its collaboration with ESI to assist tourist hotspots towards a better preparation for the foreseeable return of crowds during the post-corona recovery phase.

Education

Education has always been a logical combination for ESI with our operational consultancy work. Since 2015, ESI has been teaching the three-day "Introduction to Crowd Science" course in the Netherlands. The course was first tutored by Prof. Dr. Still, but since 2017 ESI is performing this course in the Dutch language through a licence agreement. At this time more than 150 delegates have participated in this course with many of our ex-students afterwards enrolling in the MSc program at the Manchester University (MMU). As the recent downscaling of the measures enabled the restart of these courses, we have encountered a strong demand for this course.

We have also established a special collaboration with Prof. Dr. Still and the Breda University of Applied Sciences (BUAS). Since 2019, this university has been offering a unique minor on crowd safety, in which the curriculum of the Introduction to Crowd Science course has been fully integrated. As many as 50 students each year choose this minor resulting it in being one of the most popular and highly reviewed minors on offer by the BUAS.

The students share a background in safety management, logistics, tourism or mobility. ESI performs lectures to the Dutch students and Prof. Dr. Still assesses the final works of the students in collaboration with teachers from BUAS. The third edition of this course will start next fall 2021.

Importance of Crowd Science

We have found that it is increasingly common for stakeholders to utilise the concepts of Crowd Science within the Netherlands and to do so with a more in-depth knowledge and experience. In part this is due to the efforts of ESI through the training we have provided over the years for many private organizations and local governments. More and more questions are brought to the table such as: what is the intake profile? What directions do visitors come from? Can the entrances handle the expected inflow of the crowd? What will be the behaviour of the visitors in long queues or during emergency situations? Is the event organization sufficiently prepared for such foreseeable scenarios?

The Covid pandemic has now presented the opportunity (and the need) for busy cities and leisure hotspots to use the good practices already set by the event industry. We train the professionals and support them in their analysis. Operationalising these concepts to enable overview by providing them with a good decision support matrix and enhancing their decision-making through training. We now also train the police and fire brigade with these crowd management insights and see that this positively affects their efficiency and effectivity. As a relatively small organization, we thus make a significant impact in our country soon to be packed again with leisure events, in which tourism is to be expected to bounce back rapidly out of this pandemic. This will also put an increasing pressure on maintaining the balance with liveability for the inhabitants, thus there is still plenty of work to be done!

Chapter 17

The future

> Education is when you read the fine print. Experience is what you get if you don't.
>
> Pete Seeger

On June 17th, 2021, the Manchester Evening News Arena bombing report was published, with recommendation to make the duty of care, the safety of all patrons, legislation. During the inquiry, the subject of risk assessment was scrutinised and highlights the need to improve our risk assessment methodology and understanding of crowd dynamics in built and complex spaces.

Pre-pandemic, we used to run 20–30 workshops and courses, around the world. From 2015 to 2021 we developed and supported the online MSc at Manchester Metropolitan University. For our type of work, Brexit, the University REF[1] system, and the Pandemic proved to be a perfect storm. Around the world, all major events, mass gatherings and religious festivals were suspended, with no sign of return to the old normal. At the time of writing this section, there is still uncertainty about a return to pre-pandemic events, globally. Test events (Euro 2021, Wimbledon) were indicating that "normal" would return, soon.

LEARNING DURING LOCKDOWN

As a result of the pandemic, the emphasis on understanding crowd science increased around the world. Our learners wanted to focus on the implications of crowd management in a wider range of environments and wanted to know more about the "science" that was in the spotlight, on the news and social media. Social distancing, crowd density, proximity to others, capacity in buildings and open spaces, all were under the spotlight for

1 Where professors are assessed on the number published research papers, rather than educational contribution https://www.ref.ac.uk. Our work is 90% educational, our independent research work on persistent standing was published in 2021 (after leaving MMU).

DOI: 10.1201/9781351053068-17

potential covid transmission risks. We began to see the return of major incidents[2] as the crowds came back, we saw an increased awareness on the risk assessment process, on crowd-covid safety and on the economics of events with smaller crowds.

We expanded the range of courses and now teach 100% online, around the world. Our recent partnership with the NCS[4] (USA[3]), Breda University of Applied Science (Holland) and Charles Sturt University[4] (Australia and India). We have Level 3 and Level 5 trainers supporting the courses in Holland, USA, Canada, Australia, UK and Israel and are growing our suite of online courses[5].

COVID CROWDS

At the time of writing this section, we are seeing three distinct types of crowd behaviour. The covid cautious, those individuals who are rule obeying, staying away from public gatherings and populated areas. The celebratory, those individuals (and groups) who are partying at every opportunity and the contentious crowds, those who are joining any protest to show their discontent at the situation.

As we outlined in the chapter on the decision support tools, you can model crowd behaviour in a DIM-ICE model. Table 17.1 below shows the "normal" half of the covid crowd application in the DIM-ICE matrix.

We continue to monitor and refine our models as the crowds come back, working with organisations from around the world. It is slightly ironic that we can achieve our educational aims and objectives further, faster and with more international partnerships outside of the UK academic system than we could while working within the University framework.

AFTERWORD

Diarmuid Smyth – The future of crowd science.

> Diarmuid is the CEO of Sword Security. Sword is an international provider of security, crowd management and staffing services. On the global event stage our experience ranges from Olympic to Pan Am games and from the Middle East, Europe to North and South America. Sword are dedicated to improving standards and processes in the staffing sector for the mass gathering industry.

[2] The mass fatality incident at Mt. Meron, in Israel, is currently under investigation.
[3] https://ncs4.usm.edu
[4] https://www.csu.edu.au
[5] https://crowdrisks.com/index.html

Table 17.1 Covid Crowd DIM-ICE

Covid cautious	*Ingress*	*Circulation*	*Egress*
Design			
Information			
Management			

Celebratory	*Ingress*	*Circulation*	*Egress*
Design			
Information			
Management			

Contentious	*Ingress*	*Circulation*	*Egress*
Design			
Information			
Management			

Diarmuid Smyth is the founder of the Sword Group. As an industry leader he has helped deliver events at the highest level in sports and entertainment. He has been educated in the world's leading business schools from Harvard, London School of Economics, Stern Business School, NYU and HEC Paris. He is currently undertaking a PhD in ESCP Paris.

There is nothing so practical as a good theory

In considering the future of crowd science we must remember that it is still a nascent sector. Three decades of theory, research and concept development is a relatively short period of time in the development of any academic field. The great work and foundations laid by Fruin and Still should pave the way for others within the sector to continue grow and develop it. One area of the field that needs to continue to develop is the application of the theory to its practical implementation by operators around the world. As Dr. Kurt Lewin, the social psychologist and the Founder and director of the Research Centre for Group Dynamics at Massachusetts Institute of Technology famously observed, "There is nothing so practical as a good theory".

Unfortunately, too many organisations around the world still do not apply the theories of crowd science to their events, venues, and mass gatherings. The repeated incidents, accidents, and fatalities on a regular basis are a testament to this. Too often the refrain of those on the operational front lines is "it is too academic not practical to implement on the ground." In the cold light of day in a court room it is not a robust defence to claim that crowd science frameworks and theories are too academic in the aftermath of a serious crowd incident.

A different perspective on incorporating crowd science theories into general business practices and planning is to consider it from and Environmental Social Governance (ESG) and stakeholder perspective. Much has been written on business in general on the importance of Stakeholder theory, which argues that the value creation of an organisation is interdependent with its stakeholders. This approach helps ensure that the business and its services or products are aligned with what stakeholder's value. Examples of this range from companies using fair trade products or adhering to various voluntary standards such as Carbon Footprint goals or Living Wage standards.

From a stakeholder perspective applying Crowd Science to every mass gathering while not compulsory should be a desire for every organisation. Ensuring the safety of the event and of spectators and individuals at mass gathering events is of course regulated by various health and safety standards. The strength and depth of these vary widely around the world. But Crowd Science is not just health and safety. It is easy to highlight multiple major event examples where organisers have "ticked the box" of health and safety yet failed to consider any of the fundamental aspects of crowd science in their planning. Of course, all major event organisers are focused on creating an amazing experience for their guests and they will take all reasonable steps to ensure what they believe in their eyes to be a "safe event" but unfortunately the crowd science frameworks are not incorporated within their planning. If we frame Crowd Science within an ESG approach perhaps it may make it easier for organisers, procurement departments or operational delivery teams to consider it as a *sine qua non* within their planning.

The UK is a global leader in the crowd science and crowd management sector. Countless crowd safety practitioners have gained their education and experience through UK organisations from the Emergency Planning College, the NVQ training programs to the Masters' Degree Program developed by Professor Still at Manchester Metropolitan University or the program developed by Dr. Chris Kemp at Napier University. The high standards of the UK crowd science and crowd management industry has led to UK trained professionals being in demand around the world at the highest profile events.

While this is a great reflection on the work done by UK academics and practitioners it represents both a challenge and an opportunity for the rest of the world. Breda University in Holland has developed its own crowd science program in conjunction with Professor Still and there are various other universities that are considering crowd science as a potential course for their students. The Covid Pandemic has changed the way education is conducted. The change from in person teaching to online learning has accelerated to a level that would have been inconceivable pre the Pandemic. The move to online learning should facility the easier learning of crowd science and applying its frameworks to mass gatherings around the world. Being trained or educated in the UK should no longer be the foundation for developing and implementing crowd science theories into practice. The National Centre for Sports Safety and Security (NCS[4]) based out of the University of Mississippi has in 2021 launched a Fundamentals of Crowd Science course in conjunction with Professor Still. This is of course a positive development, and the opportunities exist for other organisations and Universities around the world to work with the academic leaders to develop the theories and practices in their local countries.

Professor Still is first and foremost an individual who is focused on developing crowd science and challenging and holding to account crowd safety practices around the world. Too many practitioners are focused on protecting their local fiefdom and not sharing their knowledge. More individuals and organisations should focus on developing the field with its life saving practical implications.

As part of his commitment to develop the field Professor Still and other academics are committed to working in partnership with universities and organisations around the world. In partnering with academic and industry leaders to incorporate crowd science as a fundamental element of ESG for organisations it will, without a doubt save lives in the future. Crowd Science, its theories and frameworks are applicable to not only event organisers but for those working in mass transit, smart cities projects and cultural and religious gatherings. In the post-covid world where digitisation and online learning have become the norm, there are no geographic limitations on learning. If there is no local university offering crowd science courses, there are online options removing potential excuses from practitioners.

No one expects to attend an event and end up in hospital or lose a loved one because the organisers lacked a fundamental understanding of crowd science. Perhaps in the future there will be "Crowd Science" international standard that is developed in conjunction with University Professors and Expert Practitioners. International standards exist for numerous services, products and processes from the iconic ISO 9001 to Information Security Management, Health and Safety to Environmental Management and numerous others local and international ones. Is there an opportunity for forward thinking organisations to engage on this? Before we get to that point both academia and industry professionals need to continue to develop the science, challenge the existing approaches and question what we accept as "standard" practices. By continuing to both challenge and collaborate the industry will continue to mature not only in western countries but also though out the rest of world.

FINAL WORD

> Science isn't finished until it's communicated. The communication to wider audiences is part of the job of being a scientist, and so how you communicate is absolutely vital.
>
> Sir Mark Walport,
> *UK chief scientist.*

If you have enjoyed this book and learned something useful, something that will help you improve your crowd safety and your crowd risk analysis, please spread the word.

Index